Continuous Cultures of Cells

Volume I

Editor

Peter H. Calcott, D. Phil.
Assistant Professor
Department of Biological Sciences
Wright State University
Dayton, Ohio

CRC Press
Taylor & Francis Group
Boca Raton London New York

CRC Press is an imprint of the
Taylor & Francis Group, an **informa** business

CRC Press
Taylor & Francis Group
6000 Broken Sound Parkway NW, Suite 300
Boca Raton, FL 33487-2742

Reissued 2019 by CRC Press

A Library of Congress record exists under LC control number:

Publisher's Note
The publisher has gone to great lengths to ensure the quality of this reprint but points out that some imperfections in the original copies may be apparent.

Disclaimer
The publisher has made every effort to trace copyright holders and welcomes correspondence from those they have been unable to contact.

ISBN 13: 978-1-315-89177-4 (hbk)
ISBN 13: 978-1-351-07087-4 (ebk)

Visit the Taylor & Francis Web site at http://www.taylorandfrancis.com and the CRC Press Web site at http://www.crcpress.com

PREFACE

Continuous culture is a method used both in research and industry to grow microbes, primarily bacteria; though it has been used to grow algae, protozoa, fungi, and plant and animal cells. Continuous, or open, culture differs from the batch, or closed, culture method in so much as it protracts growth of the organisms in a time independent dimension. With this method, which is incidently more complex to operate than simple batch culture, it is possible to study more rigorously the physiology, biochemistry and genetics of microorganisms as it relates to, for instance, the influence of environmental factors.

Previous to this book, a number of reviews, symposia, and monographs have focused on continuous culture as a tool. However, no one book or symposium has managed to capture the full range of applications. This book was intended to attain this goal. The inception of the project was in 1977 when I invited some world-recognized experts in the field to contribute chapters to this treatise. The book was compiled by the summer of 1978. Thus this book documents and illustrates the knowledge known and recognized to that date. There has been, as in all rapidly advancing areas, an advance in this area since 1978 which has obviously not been included.

I have tried in this book to present as broad a perspective as possible to the subject matter. In the construction of the chapters, I have also left much up to the individual contributors. Some chapters have been written as essentially up to the minute reviews of an application or use of continuous culture while others have used data obtained in the author's own laboratory to illustrate the use of continuous culture as a problem-solving tool. Yet others have concentrated on specific topics and cited a few key ways in which continuous culture can be useful. The approach was left solely up to the contributor.

In this two-volume set, I have included chapters on the overall perspective of the technique (Chapter 1), the construction and operation of laboratory cultures (Chapter 2), and the mathematics of growth in continuous cultures (Chapter 3). Chapter 4 focuses on the use and advantage of complex and multi-stage continuous fermenters while Chapters 5 and 6 demonstrate the use of the technique in industry to produce single cell protein and fine biochemicals and drugs from simple or waste substrates. Chapter 7 demonstrates the use of the technique in studying nonsteady state or transient phenomena. Volume II also comprises seven chapters, with the first four being devoted to the application of the technique to the studies of cell metabolism, more specifically to carbon metabolism, chemical composition of cells, intermediary metabolism, and oxidative phosphorylation. Studies on the genetics of microorganisms in continuous culture is dealt with in Chapter 5. Plant cell and algae culture are focused in Chapters 6 and 7. While these two volumes cover most applications of the technique there are several which because of space were not included; these are discussed in Chapter 1.

It is hoped that this two-volume set will be useful to the established continuous culture operator as well as the researcher, teacher, and student who is interested in learning how the technique could be useful in answering both basic and applied questions in microbiology and cell biology.

Peter H. Calcott
January, 1981

THE EDITOR

Peter H. Calcott, D. Phil., is Assistant Professor of Biological Sciences at Wright State University, Dayton, Ohio.

Dr. Calcott was graduated from the University of East Anglia, Norwich, England, with a B.Sc. (Hons.) degree in Biological Sciences in 1969. He received his D. Phil. (Biology) in 1972 from the University of Sussex, England under the supervision of Professor J. R. Postgate. After 2 years postdoctoral training in the Department of Microbiology, Macdonald College of McGill University, Montreal, Quebec, Canada with Professor R. A. MacLeod, Dr. Calcott was a Professional Associate in that department from 1974 to 1976. He joined the faculty of Biological Sciences at Wright State University in 1976. Dr. Calcott has spent leaves working with Dr. D. Dean, Department of Microbiology, Ohio State University and Professor A. H. Rose, University of Bath, England.

Dr. Calcott's research interests revolve around the reaction of microbes to stress, primarily freezing and thawing and starvation. He is particularly interested in the role of cell wall and membrane structures in determining the resistance of organisms to stress, continuous culture of microbes and the role of small molecules such as cyclic AMP and cyclic GMP in cell metabolism. Dr. Calcott has published more than 50 research papers, reviews, books, and abstracts over his career.

CONTRIBUTORS

M. J. Bazin
Senior Lecturer in Microbiology
Microbiology Department
Queen Elizabeth College
London, England

Peter H. Calcott
Assistant Professor
Department of Biological Sciences
Wright State University
Dayton, Ohio

Sallie W. Chisholm
Associate Professor
Doherty Professor of Ocean Utilization
Division of Water Resources and Environmental Engineering
Civil Engineering Department
Massachusetts Institute of Technology
Cambridge, Massachusetts

F. Constabel
Senior Research Officer
Prairie Regional Laboratory
National Research Council of Canada
Saskatoon/Saskatchewan
Canada

Charles L. Cooney
Associate Professor of Biochemical Engineering
Biochemical Engineering Laboratory
Department of Nutrition and Food Science
Massachusetts Institute of Technology
Cambridge, Massachusetts

E. A. Dawes
Department Head
Reckitt Professor of Biochemistry
Department of Biochemistry
University of Hull
Hull, England

P. Doberský
Research Specialist
Department of Technical Microbiology
Czechoslovak Academy of Sciences
Praha, Czechoslovakia

J. W. Drozd
Doctor
Shell Research Limited
Shell Biosciences Laboratory
Kent, England

D. C. Ellwood
Director
Pathogenic Microbes Research Laboratory
PHLS Centre for Applied Microbiology & Research
Wiltshire, England

Ivan J. Gotham
Research Scientist I
New York State Department of Health
Albany, New York

Margareta Häggström
Doctor
Technical Microbiology Chemical Center
Lund University
Lund, Sweden

Walter P. Hempfling
Associate Professor
Department of Biology
The University of Rochester
Rochester, New York

H. Michael Koplov
Section Leader
Schering Corporation
Union, New Jersey

W. G. W. Kurz
Senior Research Officer
Prairie Regional Laboratory
National Research Council of Canada
Saskatoon/Saskatchewan
Canada

J. D. Linton
Shell Research Limited
Shell Bioscience Laboratory
Kent, England

Abdul Matin
Assistant Professor
Department of Medical Microbiology
Stanford University
Stanford, California

G-Yull Rhee
Research Scientist IV
New York State Department of Health
Albany, New York
Adjunct Associate Professor
Cornell University
Ithaca, New York

Craig W. Rice
Fellow
Department of Biochemistry and Biophysics
Division of Genetics
University of California San Francisco
San Francisco, California

J. Řičica
Senior Scientific Worker
Deputy Head
Department of Technical Microbiology
Institute of Microbiology
Czechoslovak Academy of Sciences
Praha, Czechoslovakia

A. Robinson
Head of Pertussis Vaccine Unit
PHLS Centre for Applied Microbiology & Research
Pathogenic Microbes Research Laboratory
Wiltshire, England

V. R. Srinivasan
Professor
Department of Microbiology
Louisiana State University
Baton Rouge, Louisiana

R. J. Summers
Senior Research Biologist
M. E. Pruitt Research Center
Dow Chemical, U.S.A.
Midland, Michigan

TABLE OF CONTENTS

Volume I

Chapter 1
Continuous Culture: Where it Came From and Where it is Now 1
Peter H. Calcott

Chapter 2
The Construction and Operation of Continuous Cultures.......................13
Peter H. Calcott

Chapter 3
Theory of Continuous Culture 27
M. J. Bazin

Chapter 4
Complex Systems 63
J. Řičica and P. Doberský

Chapter 5
Continuous Culture in the Fermentation Industry 97
V. R. Srinivasan and R. J. Summers

Chapter 6
Single-Cell Protein Production From Methane and Methanol in Continuous Culture 113
J. W. Drozd and J. D. Linton

Chapter 7
Transient Phenomena in Continuous Culture 143
Charles L. Cooney, H. Michael Koplov, and Margareta Häggström

Index 169

Volume II

Chapter 1
Carbon Metabolism...1
Edwin A. Dawes

Chapter 2
Bacterial Envelope Structure and Macromolecular Composition.................39
D. C. Ellwood and A. Robinson

Chapter 3
Regulation of Enzyme Synthesis as Studied in Continuous Culture...............69
Abdul Matin

Chapter 4
Microbial Bioenergetics During Continuous Culture............................99
Walter P. Hempfling and Craig W. Rice

Chapter 5
Genetic Studies Using Continuous Culture 127
Peter H. Calcott

Chapter 6
Continuous Culture of Plant Cells 141
W. G. W. Kurz and F. Constabel

Chapter 7
Use of Cyclostat Cultures to Study Phytoplankton Ecology 159
G-Yull Rhee, Ivan J. Gotham, and Sallie W. Chisholm

Index 187

Chapter 1

CONTINUOUS CULTURE: WHERE IT CAME FROM AND WHERE IT IS NOW*

Peter H. Calcott

Cultivation of cells, whether they be procaryote or eucaryote, is an important aspect of biology. Perhaps some of the most important contributions to modern biology have come from situations where cells were cultivated in vitro. Cultivation of these cells has been classically in what are termed closed systems. These closed systems can come in all sizes and shapes from the hanging drop of about 50 $\mu\ell$ to the industrial fermentor of 10,000 gal or larger. These culture systems essentially operate by the same method. A medium of required composition is maintained at the correct environmental conditions of pH, Eh, gas phase, nutrient level, and temperature. When correct conditions prevail, pregrown cells are introduced. The amount added is usually small in comparison to the final output. After a period, during which time the operator can interfere with the operation of the culture or not, the cell mass or spent growth medium is harvested — depending on what the operator is interested in.

In an industrial situation when one type of cell or product is of interest, the output of material is accomplished *in staccato* fashion. After growth has been stopped and the cells or medium harvested, the operation must be started over again. This obviously involves a period of stripping down the apparatus, cleaning, introducing new medium, and "sterilizing" of the apparatus. Clearly, if this unproductive portion could be eliminated, the efficiency of the system could be increased and hence it would be economically better. For these reasons many of the successful early attempts at continuous or semi-continuous culture methods (open systems) were investigated and developed in the 1930s and 1940s by industry. The impetus appeared to be in Europe, though North America cannot be ruled out as a major influence. Needless to say, with economics playing an important influence on the advance, the latter tended to be the result of empirical rather than truly analytical approaches.

These newer or open type systems of cultivation could be divided into truly continuous or semi-continuous types. The former would entail the maintenance of the growth medium environment with respect to temperature, pH, etc., and the subsequent introduction of cells. Once the cells had grown to the required phase, new medium would be introduced continuously with the removal of spent medium. This spent medium could be harvested with respect to cells or products. Thus, in this situation, the products or cells can be continuously produced. The semi-continuous system involves a continuous type set up, but instead of the continuous removal of material, the culture is "bled" at intervals. Once a large proportion of material is removed, fresh medium is introduced to replenish the loss. Growth then progresses before the next bleeding.

In the 1930s and 1940s great advances were made in the use of continuous culture techniques in the production of lactic acid, butanol, ethanol, yeast, and acetone. About the same time some important advances were made on the more pure scientific side of research by various workers using continuous flow type systems. Rogers and Whittier,[1] Mayer,[2] and Hadden[3] appear to have made significant contributions. In the 1950s, other advances were made in the industrial field of antibiotic, toxin, steroid, and vaccine production with the use of continuous flow systems.

Before detailed analytical approaches could advance continuous culture in the industrial and research areas, a firm foothold in the mathematics of bacterial and cell growth was needed. The pioneers in this area of research were Monod[4,5] and Novick

* This chapter was submitted in June 1978.

and Szilard.[6,7] Since the advent of these classical papers, the use of continuous culture in both research and industry has increased at a rapid rate. So much so that not only have papers been presented at national and international meetings, but whole symposiums have been devoted to the technique. It appears that most of the expansion and study of this method of cultivation of cells has been in Europe, with Czechoslovakia and Russia taking a major role in the 1950s. The 1960s saw an increase in popularity of the technique in England, particularly with Microbiological Research Establishment (M.R.E) Porton Down Group. The 1960s and 1970s have also seen the major introduction of the method to North America. This is not to say some major advances were not made before this period in the U.S. and Canada.

The topic of continuous culture has been a focal point of several symposiums and books. None of these are very recent (the last few years). However, because of the limited size of this book it is necessary to give a list of the various symposium and books available on the subject (Table 1). Each of these sources have tended to concentrate on a particular aspect of the technique and left another area weakly emphasized. Perhaps one of the major contributions to the dissemination of information is the annual review of continuous culture published in Folia Microbiolgia by such people as Ricica, Malek, and Beran. These authors have catalogued papers and abstracts, emphasizing continuous culture, published each year in the major journals into the various subject areas of biology. These obviously time-consuming reports, are invaluable tools to investigators who either want to keep up with the technique or suddenly burst into the area.

Since most symposiums or books published over the last 20 years have tended to concentrate on certain aspects of continuous culture and left other areas weakly represented, we believe this book is a timely contribution to the market, since we have attempted to bring out the major areas of biology that use continuous culture. We are sure that many readers will still express an opinion that the book is weak in certain areas In organizing this volume we have attempted to cover all major areas of interest.

Before going into detail in what is covered in the book, a major question needs to be answered. That question is "why use continuous culture when it is difficult to set up, maintain, and understand, and when the alternative is much easier". As pointed out by Tempest,[8] most papers only cite why certain methods were used; they very rarely give reasons why alternatives were not used. Thus it is impossible to ascertain why continuous culture methods have not been used more often.

Microbes, particularly bacteria, are very adaptable. They can quickly change their physiology in response to alterations in the environment by sensitive repression, derepression, and allosteric mechanisms.[9,10] Thus the activity of the bacterial cell is impossible to define apart from its environment. The classic batch culture is an example of a method of culture that rarely, if ever, maintains an environment for the cell that is constant. One could deduce that if the environment was not constant the physiology would not be constant. To understand this statement one needs to review the events associated with its growth cycle. This may seem repetitive or "old hat" to some readers, but bear with the author.

As a microbe is inoculated into a fresh growth medium its density is low and thus its ability to alter its environment is not signifcant. It then undergoes a period of adaptation to the medium, of adjustment and repair before beginning to grow. This marks the lag phase. Once the cells begin to grow and divide, the active growth phase is started. Initially a period of acceleration of growth is seen followed by a period that may last from a short portion of an hour or less to a long one of several weeks, depending on the cell type, medium, and other conditions. This period of active growth (when doubling time or growth rate is constant), is termed the logarithmic or exponential phase of growth. During the early portion of this period, the cells are still at a low

Table 1
SYMPOSIA AND BOOKS DEVOTED TO CONTINUOUS CULTURE OF CELLS

Year published	Title	Publisher
1958	*Selected Papers from 7th Int. Cong. Microbiology,* Stockholm, Sweden	Almqvist & Wiksell Stockholm, Sweden
1958	*The Dynamics of Bacterial Populations Maintained in Chemostat,* H. Moser	Carnegie Institution Publication 614 Washington, D.C. U.S.
1958	*Fermentation Kinetics and Continuous Processes 134th Symp. Am. Chem. Soc.,* Chicago, Illinois	American Chemical Society
1958	*Continuous Cultivation of Microorganisms,* edited by I. Malek	Publishing House Czechoslovakia Academie Science, Prague, Czechoslovakia
1959	*Recent Progress in Microbiology,* edited by G. Tunevall	Stockholm, Sweden
1959	*Selected papers from the Instituto Superiore de Sanita,* Vol. 2	I.S. di S
1959	*Mass Propagation of Cells, 136th Symp. Am. Chem. Soc.,* Atlantic City, New Jersey	American Chemical Society
1960	*Continuous Fermentation and Growth of Microorganisms*	Pishchepromizdat, Moscow
1961	*SCS Monograph No. 12. Continuous Cultivation of Microorganisms*	Soc. Chem. Ind. London, England
1962	*Continuous Cultivation of Microorganisms Proc. 2nd Symp., Prague*	Publishing House Czechoslovakia Academie Science, Prague, Czechoslovakia
1962	*Continuous Production of Fodder Yeast,* edited by K. Beran	SNTL, Prague, Czechoslovakia
1964	*Continuous Culture of Microorganisms,* Edited by I. Malek, K. Beran, J. Hospodka.	Academic Press, London; New York
1966	*Theoretical and Methodological Basis of Continuous Culture of Microorganisms,* Edited by I. Malek, K. Beran, Z. Fencl.	Publishing House Czechoslovakia Academie Science & Academic Press, London and New York
1967	*Continuous Culture and Microbial Physiology 4th Int. Sýmp. Continuous Culture, Porton Down, Wiltshire, England,* Edited by E. O. Powell, C. G. T. Evans, R. E. Strange, and D. W. Tempest.	Her Majesty's Stationery Office, London, England
1969	*Continuous Cultivation of Microorganisms,* Edited by I. Malek, K. Beran, Z. Fencl, V. Munk, J. Ricica, H. Smrckova.	Academic Press, London, England, and New York, U.S.
1970	*Introduction to Research with Continuous Culture,* H. E. Kubitschek.	Prentice Hall, Englewood Cliffs, New Jersey, U.S.
1972	*Environmental Control of Cell Synthesis and Function,* Edited by A. C. Dean, S. J. Pirt, and D. W. Tempest.	Academic Press, London, England, and New York, U.S.
1975	*Principles of Microbe and Cell Cultivation,* S. J. Pirt	Blackwell Press, London and Oxford, England.
1976	*Proc. 6th Int. Symp. Continuous Culture, "Applications and New Fields" Oxford, England,* Edited by A. C. R. Dean, D. C. Ellwood, C. G. T. Evans, and J. Melling	Ellis-Horwood, Chichester, England.
1976	*Yield Studies in Microorganisms, A. M. Stouthamer, Patterns of Progress in Microbiology,* PPM 3.	Meadowfield Press, Shilden Co., Durham, England.
1976	*Continuous Culture in Microbial Physiology and Ecology,* H. Veldkamp, *Pattern of Progress in Microbiology,* PPM 7.	Meadowfield Press, Shilden Co., Durham, England.
1976	*Continuous Culture Mini Symposium,* Convener A. Matin, American Society for Microbiology.	American Society for Microbiology, Washington, D.C.
1960 to 1970s	*Folia Microbiologia* — review articles by authors such as: J. Ricica, K. Beran, and I. Malek	Publishing House Czechoslovakia Academie of Science, Prague.

population density and consequently are utilizing substrates at a very slow rate and excreting waste products at a slow rate. Therefore, their environment is not changing significantly. Deductively one can conclude that their physiology is not changing. As the log phase progresses, their influence on their environment also changes. One consequence is that the population density increases to a significant proportion of the culture volume. When this occurs, the cells take up nutrients from the finite pool at a relatively high rate. In addition, they are excreting toxic products into their environment at a rapid rate, which results in a significant change in the concentration of these compounds. Clearly, when the cells are affecting their environment so drastically, their physiology would be predicted to alter at a corresponding rate. Once growth begins to slow, due to a physical limitation of space, exhaustion of a key nutrient, or to the build up of toxic material, the culture is said to enter stationary phase. This phase is far from stationary, since although cell numbers or viable count may be constant, there is an active turnover, breakdown, and resynthesis of the cell's pools and macromolecules. At this stage, the density of the culture is so large that the environment is under very tight control of the cell — and the cell by environment.

Thus, a steady state or balanced growth is only attained when the cell is not changing its environment. In batch or closed culture this would only correspond to early logarithmic phase. In microbiology, however, scientists are fond of using mid-logarithmic or stationary phase as "physiological cells", whatever the phrase means. During both these phases, the population density is significant and consequently the cell can influence its environment, which in turn can alter the cell's physiology. One can understand why mid-log and stationary phase have been used. Both these phases yield large quantities of cells from the minimum amount of medium. In a continuous culture, the microbes, usually at a high population density, are continually taking up the "last trace" of limiting nutrient and excreting toxic metabolites. But when a steady state is attained, the rates of absorption of limiting nutrient, excretion of toxic products, cell growth, and loss of cells from the reactor are balanced and constant with time. Thus the environment of the cell is constant. This constant environment leads then to a constant physiology.

How then can alterations in physiology be studied and controlled in batch culture? One such parameter that can be studied is alteration in growth rate. In batch culture, growth rate is governed by such factors as complexity of the medium, various physical and chemical factors such as pH, Eh, temperature, gas phase, and the intrinsic maximum specific growth rate (μ_{max}) of the cell line. To alter the growth rate of a batch culture requires manipulating these factors. However, altering these factors obviously alters the cell's environment. If one alters the environment, the physiology is altered. One then asks the question, "Has the physiology of the cell been altered by an alteration of the growth rate of the culture or by some other factor such as temperature, population density, or medium complexity?" Clearly, it would be advantageous to divorce these two sets of factors. Continuous culture lends itself to this problem admirably. In continuous culture, it is possible to attain steady-state conditions in a culture at defined and constant growth rate when such factors as medium complexity, temperature, and pH are held constant. Thus it is possible to investigate the role of growth rate on physiology without alterations in medium complexity and temperature confounding the situation. In a like manner, it is possible to investigate the physiological changes of an organism imposed by alterations in temperature, keeping growth rate and medium complexity constant. Similarly, the effect of medium composition on physiology can be studied at constant temperature and growth rate. Thus continuous culture can be used to maintain cells at constant defined growth conditions and thus maintain their physiology. This type of approach has been used by many to obtain physiologically reproducible and comparable populations. To illustrate the situation, an example is now cited.

Lee[11] compared the survival after freezing and thawing of several bacterial species in samples of soil obtained from local areas (Montreal, Canada), Inuvik (North West Territories, Canada), and Churchill (Manitoba, Canada) after freezing and thawing. To obtain reliable and comparable populations needed for survival studies he grew the bacteria in a simple defined minimal medium at 37°C under carbon-limitation in chemostat. With these populations, he obtained very reliable constant populations; inter-experiment variation was maintained at a low level. In addition the response of the organisms could be compared, since they were grown under very similar conditions of nutrient limitation medium complexity, temperature, population density, and growth rate.[12]

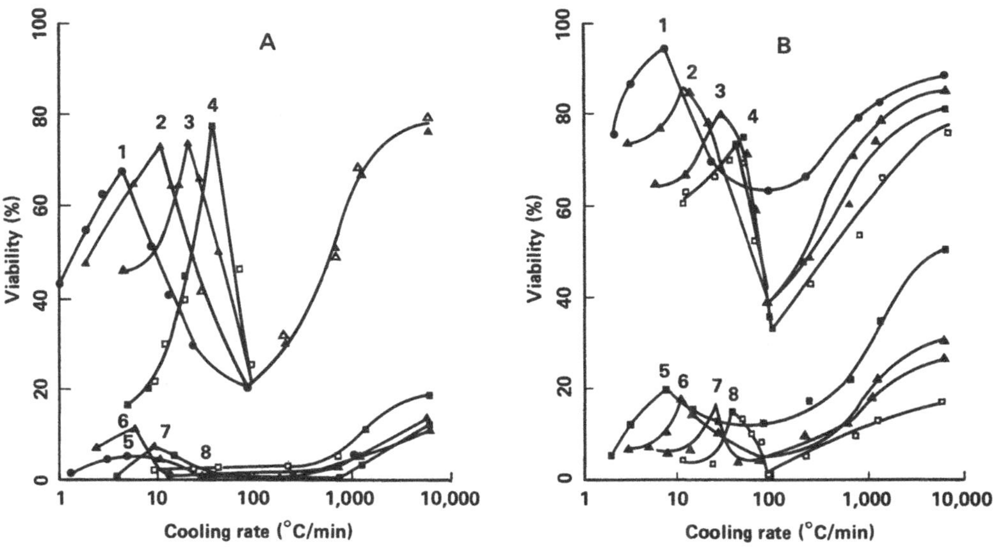

FIGURE 1. Effect of growth rate on crysurvival of glycerol and ammonium limited *E. coli* frozen in distilled water or saline. Twice washed cells of *E. coli* grown in chemostat under either glycerol-limited (A) or ammonium-limited (B) at various growth rates were cooled at various rates and warmed rapidly. Viability was determined by slide culture. Growth rates A1 and A5 μ = 0.10 h^{-1}; A2 and A6, μ = 0.20; A3 and A7, μ = 0.39; A4 and A8, μ = 0.60; B1 and B5, μ = 0.11; B2 and B6, μ = 0.218; B3 and B7, μ = 0.40; B4 and B8, μ = 0.54. Freezing menstrua were 1 to 4 distilled water, curves 5 to 8, 0.85% NaCl. (Reproduced with permission of the National Research Council of Canada from Calcott, P. H. and MacLeod, R. A., *Can. J. Microbiol.*, 20, 683, 1974.)

Another advantage of continuous culture is that transient type phenomenon can be "trapped" and reproduced continuously. An example of this is of production of carbohydrate reserve material. It is known that enteric bacteria, when grown in an ammonium-limited medium, produce carbohydrate material once the log-stationary transition has passed. Similar cells grown in carbon-limited conditions do not show this response and build up little or no carbohydrate reserve material. By growing enteric bacteria such as *Escherichia coli* under carbon- or nitrogen- limited conditions, Calcott and MacLeod[13] were able to determine the role of carbohydrate reserves in cryoresistance. They grew populations at growth rates between 0.1 and 0.64 h^{-1} and determined their gross cellular composition (carbohydrate, mean cell mass, protein, and RNA) and their cryoresistance in water and physiological saline at a variety of cooling rates ranging from 1 to 6000°C $min.^{-1}$ They concluded that cryoresistance related directly to carbohydrate content. Cells grown under conditions that allowed synthesis of large levels of carbohydrate showed higher cryoresistance under all conditions than cells that contained lower levels of the polymer (Figures 1—3).

Sometimes, situations arise in continuous culture that would never occur in batch culture. These situations offer interesting analytical problems. One example of this is the situation of cultures of enteric bacteria that produce hyper levels of the enzyme β-galactosidase. In continuous culture after at least 30 to 40 generations the cells produce, only under lactose-limitation, levels of the enzymes between 10- and 40-fold higher than comparable cultures grown in batch culture. This situation is dealt with in more detail in the chapter on genetics in continuous culture (by Calcott).

Another interesting example of situations that occur in continuous culture alone is aging in bacteria grown in chemostat. Postgate and Hunter[14] were the first to demon-

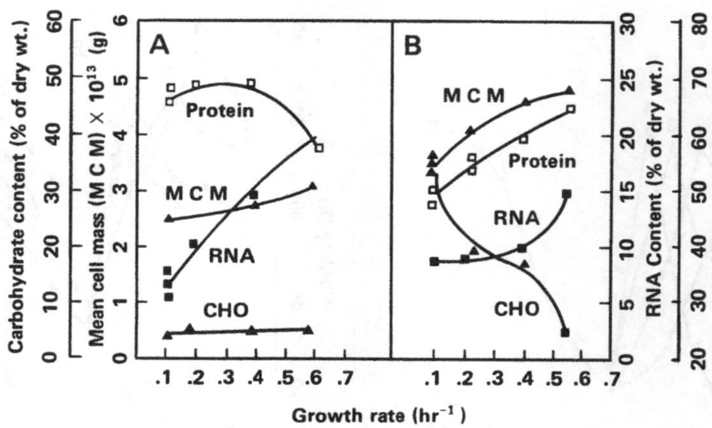

FIGURE 2. Variation of mean cell mass, RNA, protein, and carbohydrate contents of *E. coli* with growth rate and nutritional status. Steady-state population of glycerol-limited (A) or ammonium-limited (B) *E. coli* from chemostat populations were assayed for mean cell mass (MCM), protein, RNA, and carbohydrate (CHO) contents of each growth rate. (Reproduced with permission of the National Research Council of Canada from Calcott, P. H. and MacLeod, R. A., *Can. J. Microbiol.*, 20, 683, 1974.)

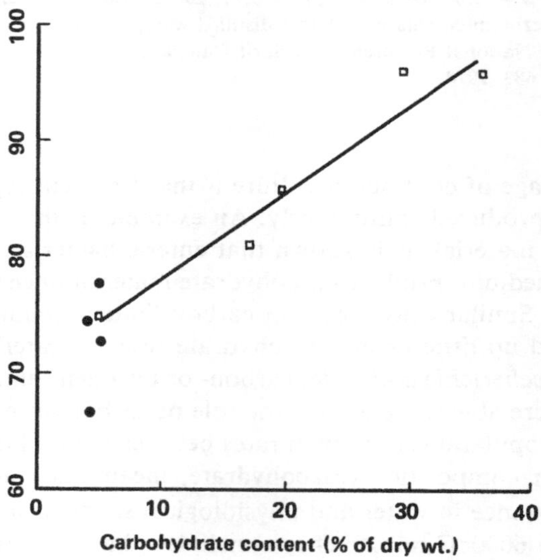

FIGURE 3. Correlation between carbohydrate content and cryosurvival of glycerol and ammonium-limited *E. coli*. Viability from freezing and thawing in distilled water at the optimum for survival and the carbobydrate content of glycerol limited (•) and ammonium-limited (□) cells were determined as in Figures 1 and 2. (Reproduced with permission of the National Research Council of Canada from Calcott, P. H. and MacLeod, R. A., *Can. J. Microbiol.*, 20, 683, 1974.)

strate viabilities of less than 100% in actively growing cells. They showed that it was possible to cultivate *Klebsiella aerogenes* in chemostat at various nutrient limitations at progressively lower and lower growth rates (Figure 4). In particular, carbon-limited

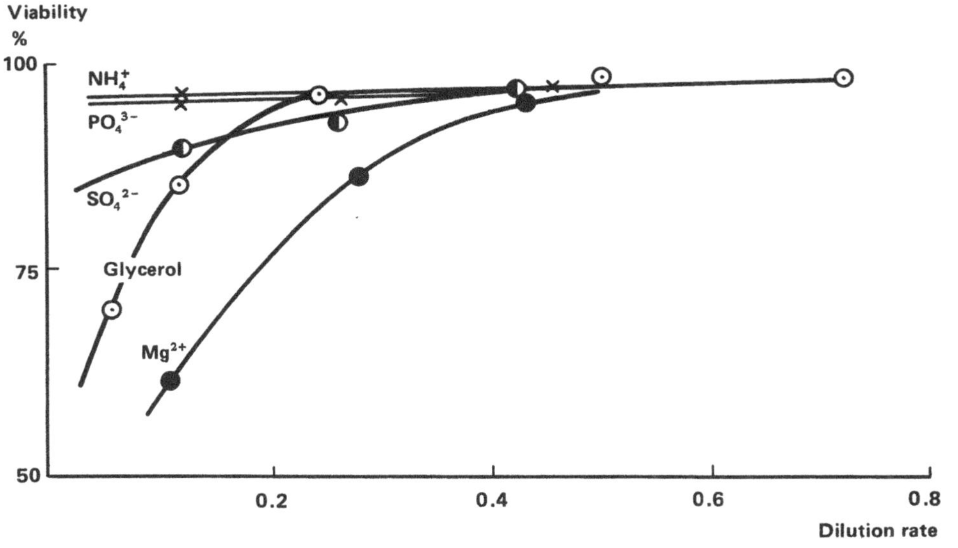

FIGURE 4. Effect of nutritional status on the steady-state viability of chemostat cultures of *Klebsiella aerogenes*. Data plotted from Table 4 of Postgate and Hunter (1962). Viability was determined by slide culture on population of *Klebsiella aerogenes* maintained in chemostat at various growth rates and under various nutrient limitations. Nutrients that limited growth are indicated on the curves. (Reproduced with permission of the Society of General Microbiology from Postgate, J. R. and Hunter, J. R., *J. Gen. Microbiol.*, 29, 233, 1962.)

organisms grown at low dilution rates were moribund, i.e., they contained both living and dead cells, as measured by slide culture on a complex medium. Under these conditions the growth rate of the viable portion in the chemostat would be higher than the apparatus as a whole. Although there should have been a powerful selection pressure in favor of resistant organisms, these authors did not detect any change in the survival properties of the population.[15]

Tempest et al.[16] continued the studies and gave detailes of the polymer content, morphology, and other properties of moribund populations of carbon and nitrogen-limited populations of the organism. These populations showed unusual snake-like morphology with an aberrant metabolism. Yield coefficient and respiration decreased with decreasing growth rate, whereas carbohydrate and RNA contents showed no obvious patterns at low dilution rates. Since ammonia and carbon-limited populations showed this response, they argued that the moribund state was not due to "starvation" for carbon source to supply the maintenance energy. The results of these workers demonstrated a μ_{min} of 0.009hr^{-1} or a doubling time of approximately 77 hr.

The concept of μ_{max} is an accepted parameter Maynard-Smith defines as being due to a queuing problem for protein synthesis.[17] With various assumptions he has even attempted to explain why higher eucaryotes have a slower μ_{max} than bacteria. The concept of μ_{min} (the minimum growth rate) is not equally accepted, however. This concept is challenged by several workers, notably Pirt[18] who believes the low viabilities are nutritional artifacts. He suggests that the shock of transfer from a nutrient limiting situation in the chemostat to a nutritionally complete (nutrient agar supplemented with yeast extract and a meat digest) was too dramatic and a proportion of the culture died. This suggestion or explanation has never been tested. Another criticism by Kubitschek[19] is that the bubbles or shear forces imposed on the culture are responsible for killing the culture. As the growth rate is lowered the time of residence of the culture and the time spent in contact with the shearing force is increased. Presumably this inactivation or killing is time dependent. Again this idea has not been tested.

Sinclair and Topiwala[20] have proposed a mathematical model of the relationship between viability and dilution rate in the chemostat. The model has no biochemical basis, but fits the data of Postgate and Hunter[14] and Tempest et al.[16] Certain of the assumptions are to be questioned: dead and live organisms have equal mass, and while live organisms actively metabolize, dead ones are metabolically inactive. At least one of these has been shown to be incorrect.[14] Clearly, the concept of μ_{min} and the objections to it need to be answered. This area of microbiology is apparently devoid of any workers at the present and appears to be so in the future. Several questions need answering:

1. Is the low viability of slowly grown populations an artifact of viability determination media?
2. Do shearing forces play a role in these low viabilities?
3. If low viabilities are real, do the dead portions resemble the living or not in such factors as DNA, RNA contents, and macromolecular synthesis potential?
4. Are the cellular control mechanisms functioning properly?

Another advantage of continuous culture is that the technique allows competition studies to be conducted. The types of interactions that can be studied include intraspecies wild-type vs. mutant and interspecies bacteria-bacteria competitions. In addition predator-prey type interaction such as between bacteria and protozoa, bdellovibrio and bacteria, bacteria and viruses can be studied. These will be dealt with in more detail in a chapter by Calcott, and in a book by Veldkamp.[21]

Certain considerations are worth mentioning. Batch culture methods of growth always saturate the cell's enzyme systems with substrates at concentrations several orders of magnitude higher than the K_s. The only time they will even attain a situation of limitation is when the cells are entering stationary phase. Consequently, cells in steady state in batch culture are always in contact with high levels of nutrient. Thus if two organisms, competing for the same substrates, are present the winner will be the cell type that can utilize the substrate the most rapidly and attain the highest growth rate. This situation is not that found in nature. As outlined by Tempest,[8] "rarely are organisms isolated from one another in a closed or protected environment, and rarely would nutrients be in such an excess to maintain the highest growth rate". Logically, then batch culture and nature are two totally different situations. This was so clearly pointed out by Kluyver when he stated that all pure cultures are laboratory artifacts (quoted by Postgate[22]). Continuous cultures, however, are run so that a low concentration of a limiting nutrient controls the growth rate of the population. Routinely, the concentration of the nutrient will be below or just about the K_s of the system. For this reason, growth in continuous culture is maintained at sub-maximal rates. The rates can, in fact, be maintained at those found in nature. Continuous culture growth can be likened to growth in nature and many workers have used this premise to study ecological problems in chemostat (see review by Veldkamp[21]).

The work of Maaløe and Kjeldegaard[10] was a major breakthrough in the understanding of control of macromolecular synthesis. They studied the control of these molecules by their shift-up and shift-down type approaches. Basically, a batch culture at steady state was shifted up to a growth situation that increased the growth rate or shifted down to a growth situation that decreased the rate. This was accomplished by altering the complexity of the medium and/or temperature, etc. They followed the differential rate of synthesis of cell number, mass, DNA, RNA, and protein contents. These types of approach pioneered the understanding of the relative control of the synthesis of the molecules. Since to increase growth rate in batch cultures requires a concomitant change in temperature and medium complexity, the precise meaning of a shift-up is difficult to interpret. It should be clear to the reader that a simple shift-up experiment can be performed using a chemostat. A steady state is attained at a low dilution rate and the cell content of various macromolecules is determined. The pump is increased to a new value and this then increases the flow of nutrients to the reaction vessel. In response to this increase in nutrient level, the organisms are posed with an environment that can maintain a higher growth rate. The rate of change of these macromolecules can be followed until a new steady state is attained. The reader should be aware that the shift-up has involved an increase in growth rate with no alteration in medium composition or temperature. These approaches have been utilized by many workers to advantage.

This book has attempted to take some of the areas of biology and document approaches involving continuous culture that have been beneficial to their study. Some chapters have been presented as reviews of the field, containing a relatively exhaustive literature survey. Others have concentrated on specific areas and cited a few experiments to demonstrate how continuous culture can be beneficial. Still others have relied on their own published works and shown how a problem can be solved by the use of technique. The choice of approach has been left entirely to the contributor.

In these volumes we have included chapters with an overall perspective on the use of continuous culture (this chapter), the construction and operation of simple continuous cultures (Chapter 2 of Volume I), the mathematics of continuous culture and growth (Chapter 3 of Volume I), and complex systems, their operation, construction, and mathematics (Chapter 4 of Volume I). These introductory chapters serve a function to furnish the reader with the actual feel of the potential and operation of the instrument, both the simple, popular one stage system and the more complex system with two and more reaction vessels. The next group of chapters deals with ways in which cells, primarily bacteria, can alter their composition (Chapter 2 of Volume II), their intermediary (Chapter 3 of Volume II), and carbon metabolisms (Chapter 1 of Volume II) in response to subtle changes in the environment. In addition a chapter on how the efficiency of utilization of substances can be determined easily in bacteria with the technique is included (Chapter 4 of Volume II).

Competition between organisms is reviewed in Chapter 5 of Volume II emphasizing genetic studies using continuous culture.

Bacteria are not the only cell type to be cultured by continuous techniques. It is also possible to culture plant cells (Chapter 6 of Volume II). In particular, aspects of construction of specialized culture apparatus and their mathematical treatment is dealt with in detail in this chapter. Originally, conceived that filamentous organisms would not grow in continuous culture systems ideally, it is being found that the cultivation of some of these is possible. One cell type capable of being cultivated in this manner is algae (Chapter 7 of Volume II). The final two chapters focus on applications in industry with chapters dealing with the continuous production of single cell protein (Chapter 6 of Volume I) and on other industrial fermentations for biochemicals, drugs, and organic chemicals (Chapter 5 of Volume I).

Scholars of continuous culture will immediately recognize areas of study totally ignored in the text. Some areas of obvious omission are the growth of spore forming bacteria[23,24] and fungi[25-28] in continuous culture. This technique has been used admirably to investigate the developmental stages triggering sporulation. In addition, the total lack of study of the fungi and protozoa in continuous culture is evident. The use of continuous culture to study stress on the physiology of organisms is also not included. Some early work by Harder and Veldkamp[29] was directed to the impairment of protein synthesis in an obligately psychrophilic *Pseudomonas* sp. grown at super-optimal temperatures. This is also a very promising, but unexploited area.

Likewise, the omission of the lower animals in continuous culture is striking, though the author can only defend himself by stating that this book is on continuous culture of cells and not animals or aggregates of cells.

REFERENCES

1. **Rogers, L. A. and Whittier, E. V.,** The growth of bacteria in continuous flow broth, *J. Bacteriol.,* 20, 126, 1930.
2. **Mayer, H. V.,** A continuous method of culturing bacteria for chemical study, *J. Bacteriol.,* 18, 59, 1929.
3. **Hadden, E. C.,** Apparatus for obtaining continuous bacterial growth, *Trans. R. Soc. Trop. Med. Hyg.,* 21, 299, 1928.
4. **Monod, J.,** *Recherches sur la Croissance des Cultures Bacteriennes,* Hermann and Cie, Paris, 1942.
5. **Monod, J.,** La technique de culture continue: theorie et applications, *Ann. Inst. Pasteur (Paris),* 79, 390, 1950.
6. **Novick, A. and Szilard, L.,** Experiments with the chemostat on spontaneous mutation of bacteria, *Proc. Natl. Acad. Sci. U.S.A.,* 37, 708, 1950.
7. **Novick, A. and Szilard, L.,** Description of the chemostat, *Science,* 112, 715, 1950.
8. **Tempest, D. W.,** The place of continuous culture in microbiological research, *Adv. Microb. Physiol.,* 4, 223, 1970.
9. **Herbert, D.,** 1961. The chemical composition of microorganisms as a function of their environment, in *Microbial Reaction to Environment,* Meynell, G. G. and Gooder, H., Eds., Cambridge University Press, Cambridge, 1961, 341.
10. **Maaløe, O. and Kjeldegaard, N. O.,** *Control of Macromolecular Synthesis: a Study of DNA, RNA, and Protein Synthesis in Bacteria,* W. A. Benjamin, New York, 1966.
11. **Lee, S. K.,** The survival of bacteria in different types of Arctic soil, *Can. Fed. Biol. Sci.,* 19, 187, 1976.
12. **Calcott, P. H.,** Nutritional factors affecting microbial reaction to stress, in *CRC Handbook of Nutrition and Food,* Richcigl, M., Ed., CRC Press, Cleveland, Ohio, 1978.
13. **Calcott, P. H. and MacLeod, R. A.,** Survival of *Escherichia coli* from freeze-thaw damage: influences of nutritional status and growth rate, *Can. J. Microbiol.,* 20, 683, 1974.
14. **Postgate, J. R. and Hunter, J. R.,** The survival of starved bacteria, *J. Gen. Microbiol.,* 29, 233, 1962.
15. **Postgate, J. R.,** The viability of very slow growing populations: a model for the natural ecosystems, *Bull. Ecol. Res. Comm.,* (Stockholm), 17, 287, 1973.
16. **Tempest, D. W., Herbert, D., and Phipps, P. J.,** Studies on the growth of *Aerobacter aerogenes* at low dilution rates in a chemostat, in *Microbial Physiology and Continuous Culture,* Powell, E. O., et al. Eds., Her Majesty's Stationery Office, London, 1967.
17. **Maynard-Smith, J.,** Limitation on growth rate in *Microbial Growth,* Meadow, P. M. and Pirt, S. J., Eds., Cambridge University Press, 1969, 1.
18. **Pirt, S. J.,** Introductory lecture: prospects and problems in continuous flow culture of microorganisms, in *Environmental Control of Cell Synthesis and Function,* Dean, A. C. R., Ed., Academic Press, London, 1972.
19. **Kubitschek, H. E.,** *Introduction to Research with Continuous Culture,* Prentice Hall, Englewood Cliffs, N.J., 1970, 50.
20. **Sinclair, I. and Topiwala, T.,** Model for viability and continuous culture, *Biotechnol. Bioeng.,* 12, 1069, 1970.
21. **Veldkamp, H.,** Continuous culture in microbiology, physiology, and ecology, *Patterns of Progress in Microbiology,* PPM 7 Merrow Publishing, Durham, England, 1976.
22. **Postgate, J. R.,** *Microbes and Men,* Penguin Publishing, Harmondsworth, England, 1969.
23. **Ricica, J.,** Sporulation of *Bacillus cereus* in multistage continuous culture, in *Continuous Culture of Microorganisms,* Malek, I., Beran, K., Fencl, Z., Munk, V., Ricica, J., Smrckova, H., Eds., Academic Press, London, 1969, 163.

24. **Dawes, I. W. and Mandelstam, J.**, 1969. Biochemistry of sporulation of *Bacillus subtilis*, 168: continuous culture studies, in *Continuous Cultivation of Microorganisms*, Malek, I., Beran, K., Bencl, Z., Munk, V., Ricica, J., and Smrckova, H., Eds., Academic Press, London, 1969, 157.
25. **Larmour, R. and Marchant, R.**, The induction of conidiation in *Fusarium culmorum* grown in continuous culture, *J. Gen. Microbiol.*, 99, 49, 1977.
26. **Ng, A. M. L., Smith, J. E., and McIntosh, A. F.**, Conidiation of *Aspergillus niger* in continuous culture, *Arch. Mikrobiol.*, 88, 119, 1973.
27. **Righelato, R. C., Trinci, A. P. J., Pirt, S. J., and Peat, A.**, The influence of maintenance energy and growth rate on the metabolic activity, morphology and conidiation of *Penicillin chrysogenium*, *J. Gen. Microbiol.*, 50, 399, 1968.
28. **Sikyta, B., Slezak, J., and Herold, M.**, Growth of *Streptomyces aereofaciens* in continuous culture, *Appl. Microbiol.*, 8, 233, 1961.
29. **Harder, W. and Veldkamp, H.**, Impairment of protein synthesis in an obligately psychrophilic pseudomonas sp. grown at superoptimal temperatures, in *Continuous Cultivation of Microorganisms*, Malek, I., Beran, K., Fencl, Z., Munk, V., Ricica, J., and Smrckova, H., Eds., Academic Press, London, 1969, 59.

Chapter 2

THE CONSTRUCTION AND OPERATION OF CONTINUOUS CULTURES*

Peter H. Calcott

TABLE OF CONTENTS

I. Basic Concepts 14

II. Types of Continuous Culture 14

III. Components of Continuous Cultures 17
- A. Nutrient Supply 17
- B. Growth Vessel 18
- C. Mixing and Aeration 19
- D. Measuring Culture Absorbance 19
- E. Spent Medium and Cells 20

IV. Operation 20
- A. Media 20
- B. Sterilization 21
- C. Inoculation 21
- D. Sampling 21

V. Specialized Situations 22
- A. Anaerobic Cultures 22
- B. High Pressure Systems 22
- C. Slow Growing Cultures 22
- D. Pathogenic Organisms in Continuous Culture 22
- E. Industrial and Complex Continuous Cultures 23

VI. Complications in the Use of Continuous Culture 23
- A. Contamination from Outside 23
- B. Contamination from Inside 23
- C. Foaming 24
- D. Wall Growth 24
- E. Failure of the Inoculum to Grow 24
- F. Growth of Cells in the Medium Inlet Line 25
- G. Mechanical and Electrical Failures 25

VIII. Commercially Available Apparatus 25

References 26

* This chapter was submitted in June 1978.

I. BASIC CONCEPTS

The purpose of this chapter is to describe in detail the basic make up and variations in construction of continuous cultures that the researcher can use. In addition, the various advantages and disadvantages of each type of culture will be described.

All continuous cultures comprise three basic units: (1) a growth medium reservoir, where the sterilized medium is stored before being delivered to, (2) the culture vessel or reactor, where the cells metabolize the nutrients. At a defined rate, spent medium and cells are removed from the reactor and pass into, (3) the spent medium reservoir. This basic construction is outlined in Figure 1.

In addition to these basic parts, certain criteria have to be met to allow the culture to operate at steady-state conditions (see chapter by M. Bazin on the meaning of this):

1. The volume must remain constant. A method of removing biomass and spent medium at the rate of new medium introduction is required.
2. The flow of new medium must be controlled and be able to be kept constant.
3. The reactor contents must be maintained as homogenous as possible. That is, when new medium is introduced, there must be rapid and complete mixing of the contents.

This basic continuous culture can be constructed in the laboratory very easily with readily available glassware, and can be used in many simple experiments where high population densities and yields of cells are not required. A very simple continuous culture of this sort was devised and used successfully by Novick and Szilard.[1]

Often, however, larger quantities of cells are needed for certain biochemical and physiological experiments. Under these circumstances more elaborate arrangements are necessary. A stylized inexpensive continuous culture is illustrated in Figure 2. This can be constructed in the laboratory[2] or bought commercially[3] (Figure 3A and B).

II. TYPES OF CONTINUOUS CULTURE

Before continuing into a description of the details of each component in a continuous culture, it is necessary to digress and describe the various types of continuous cultures. Continuous cultures can be divided into three types depending on the method employed to attain steady states. The simplest, the chemostat, relies on a set constant flow rate (F), which together with a constant volume (V) defines the dilution rate ($D = F/V$), which under steady-state conditions is equal to the growth rate of the population (μ). Thus, the operator defines the growth rate of the population by manually adjusting the flow rate of the fresh medium into the culture vessel. The population density is controlled by the concentration of limiting nutrient in the inflowing medium, and consequently, controlled directly by the operator. Routinely, the dilution rate is set at some value less than the maximum growth rate (μ_{max}) of the population in the particular medium. Attempts at running the culture at D values very close to the μ_{max} often result in "wash out" of the cells, due to fluctuations in the flow rate and dilution rate. If growth rates near μ_{max} are required, the operator often uses an alternative form of continuous culture, the turbidostat. These cultures maintain steady states by the continual monitoring of optical density or turbidity. In this manner, as the culture grows the optical density or turbidity increases, and the amount of light passing through the culture decreases. This change is detected by a photoelectric detector. Once a critical value is attained, fresh medium is allowed to enter, which dilutes the culture. This results in a decrease in turbidity or optical density that stops the entry

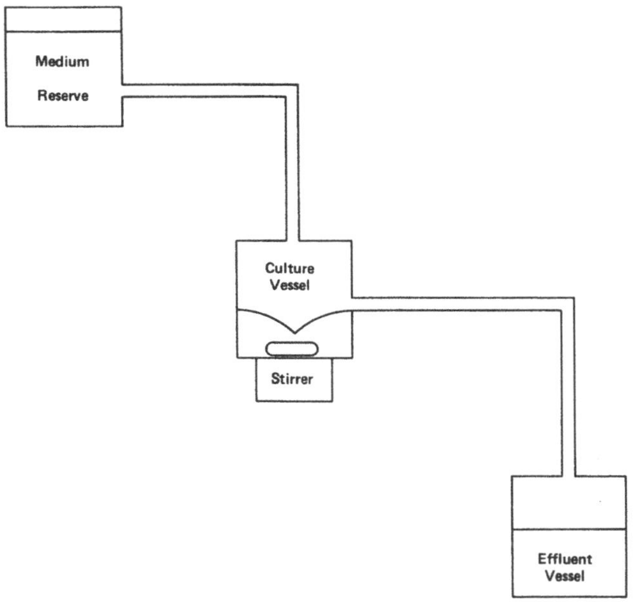

FIGURE 1. A stylized continuous culture showing the basic components.

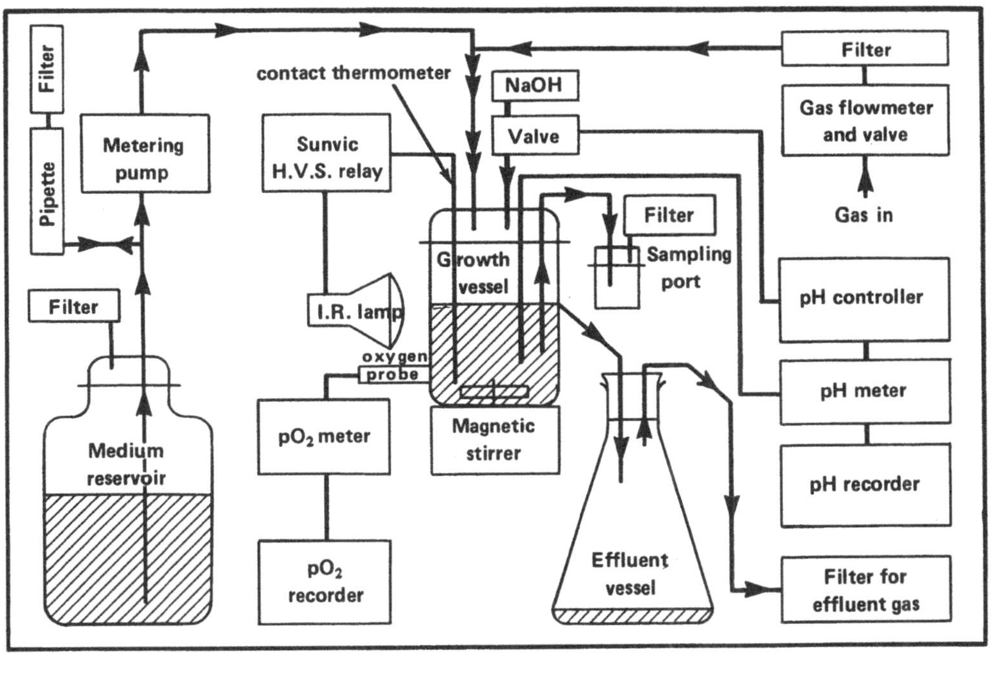

FIGURE 2. A schematic diagram of a continuous culture apparatus. (Taken with permission from Baker, K., *Lab. Pract.*, 17, 817, 1968, figure 1.)

of fresh medium. Spent medium is also allowed to flow out to maintain constant volume.

It is widely accepted that cells grown in chemostats are under external control, while those grown in turbidostats are under internal control.

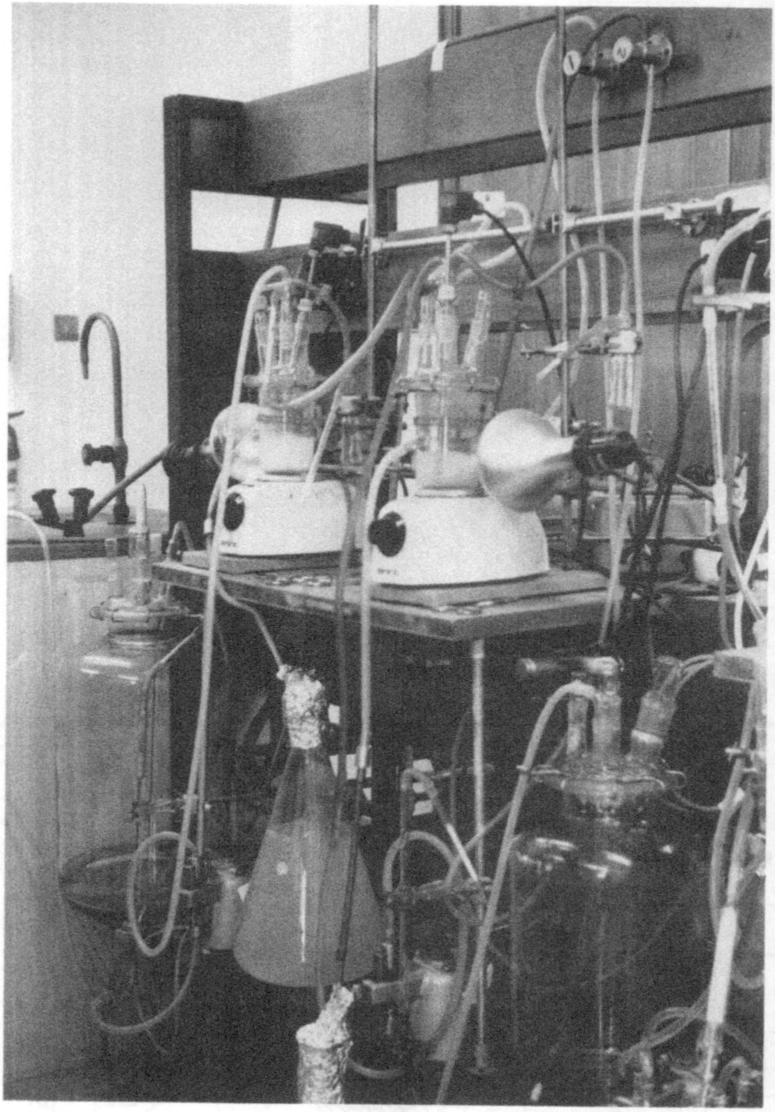

A

FIGURE 3. Chemostats at work. (A) A homemade pair of chemostats used by the author in the laboratory of J. R. Postgate, ARC Unit of Nitrogen Fixation, University of Sussex, England. (B) A commercially available chemostat manufactured by New Brunswick and Co. (Picture used with permission of New Brunswick and Co.)

Product or substrate controlled continuous cultures are like turbidostats in that the cells grown in them are under internal control, since the rate of flow of fresh medium is controlled by the utilization of certain substrates or production of some products of metabolism. Examples of these substrates or products are sulfide, potassium, pH, carbon dioxide, or those whose rapid measurement can be performed automatically.

B

III. COMPONENTS OF CONTINUOUS CULTURES

Since the surface areas of continuous cultures that come into contact with growth medium are much larger than batch cultures, care must be taken so that the material is not toxic or does not release toxic substances. Routinely, glass, perspex, stainless steel, "Teflon"®, and certain types of silicone rubber tubing are preferred materials.

A. Nutrient Supply

The components responsible for nutrient supply can be divided into two parts, a reservoir to maintain sterile medium and a device to meter the medium to the culture vessel. Virtually any vessel can be used to hold and sterilize the medium. Most often, 10 or 20 ℓ glass carboys can be used, so long as a vent to the atmosphere (properly sealed with cotton or glass wool to prevent contamination) is maintained so that medium can flow from the vessel without setting up a partial vacuum. The size of the vessel is defined by the size of the culture vessel, the desired maximum dilution rate, the size of available autoclaves, and the operator's desire to be a medium preparation technician.

The methods used to control flow rate are diverse. The two commonly used are

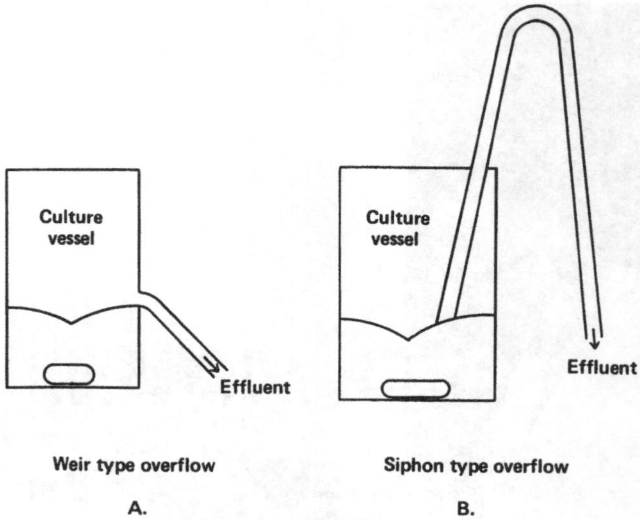

FIGURE 4. Two main types of overflow mechanisms used in continuous cultures. (A) Weir type overflow. (B) Siphon-type overflow

pumps and capillary tubing. In the former, the medium should only come into contact with the tubing and thus avoid sterilization of whole pumps. By varying the tubing internal diameter and the speed of the pump, a wide range of flow rates can be attained. The type of pump can be either the peristaltic rotary or finger type, or alternatively a "T" type pump can be used.[2] The capillary type control method is dependent on the resistance of flow of liquids through the tubing. Coarse control of the flow rate is provided by varying the internal diameter and length of the tubing. Fine control is accomplished by varying the pressure differential, itself defined by the vertical distance between the air inlet in the reservoir and the point where liquid is added to the culture.

In addition, it is necessary to know precisely the flow rate (F) of medium into the culture vessel to calculate growth rate (μ). Either the flow rate of the pump or capillary tubing is ascertained before the culture is sterilized or a device is incorporated into the inflow medium line before the metering device. Perhaps the simplest is a pipette on a branch line (Figure 3).

B. Growth Vessel

Growth vessels are usually made of all glass or perspex[4] or a combination of glass and stainless steel.[5] The size of the growth vessel can be varied from 20 mℓ to several liters, depending on the operator's desires.

Removal of spent medium and cells to maintain culture volume is accomplished by two main methods. One is a weir type with the exit tube leaving the vessel on the side at the desired height (Figure 4). Alternatively, a siphon type can be used, frequently relying on the positive pressure of the inflowing gas phase to remove liquid (Figure 4).

Inflow medium is routinely added to the vessel in a drop wise fashion. In addition, it is necessary to prevent contamination of the inflow medium from organisms in the culture vessel. Many variations on the simplest Pasteur pipette exist. Often the gas phase is passed by this to prevent aerosolized organisms from establishing in the line. The inflow gas is sterilized by passage through appropriate membrane filters, cotton, or glass wool. In addition, the air should be moistened to prevent undue evaporation of the culture volume. This can be performed with a simple Dreschel bottle in the line.

A method of measuring air flow rate is often desirable. A simple gas flow meter is both effective and cheap.

The culture vessel should also be equipped with ports at the top or side to take a sampling device. These allow the rapid withdrawal of material from the culture without interfering with its functioning.

Other ports can be used to accommodate, pH electrodes, O_2 probes, specific ion probes, thermometers, so that continual monitoring of the culture is possible. In addition, certain properties of the culture often need not only be monitored, but controlling elements for such parameters as pH, O_2 tension, and temperature are needed. In all of these systems, three parts are recognized; detectors, control boxes, and effectors. The detectors come into intimate contact with the culture, so they should be capable of being sterilized (preferably by autoclaving) and be separable from the control boxes. The control boxes are beyond the technical knowledge of the author, but many are commercially available directly from manufacturers or from supply houses. Alternatively, easily constructed ones are described adequately in the literature.[2,5] The effectors include the following types: oxygen tension is controlled by increasing or decreasing the air or oxygen flow to the culture; pH control is attained by the controlled addition of concentrated alkali or acid to the culture using mostly solenoid activated valves.[2,5] These liquids need not routinely be sterilized, since survival of organisms in these environments is not significant. However, for dense cultures — or where metabolism is high when considerable quantities of acid or alkali are needed to be added, the rate of flow of these components should be considered when calculating the dilution rate. Temperature control can be effected by a "hot finger" immersion type heater or an infrared lamp. The former needs to be able to be autoclaved separate from the control box. Frequently, temperature control can be accomplished by a constant temperature room or water bath.

C. Mixing and Aeration

Rapid efficient mixing of the culture can be performed by two major methods: a stirring impeller or magnetic bar. The former requires a drive motor above the culture with an entry port for the drive rod. This junction needs to have a bacteria-proof gland. This has been successfully employed in several systems.[5] The second type is accomplished with a "Teflon"® or PTFE-coated stirring bar driven by a magnetic stirrer. If the culture is temperature controlled in a constant temperature room, care should be taken to ensure the heat of the motor does not raise the culture temperature above that of the room. Many stirrers are commercially available that do "not" heat the solution (the heat escape is maximized). Since the advent of commercially available water baths with stirrers situated beneath or immersible stirrers, it has been possible to use a stirring bar with a water bath arrangement. Since stirring is used to not only mix the culture, but aid in the transfer of oxygen to solution, many impellers have internal channels with holes to allow escape of air directly into the culture.

In certain situations baffles are added to the culture vessel to increase the mixing properties. In most laboratory cases, the simpler stirring bars are adequate to aerate cultures. The author has successfully grown *Azotobacter* (known for its high respiration rate) in a culture stirred with 1½ in. bar and maintained at least 95% saturation in the liquid phase.[14]

D. Measuring Culture Absorbance

This is dealt with separately, since it is an integral part of only turbidostats. Routinely, culture is circulated from the main vessel through an optical cell to a photometer (Figure 5). This is routinely done with a peristaltic pump. After measurement the cul-

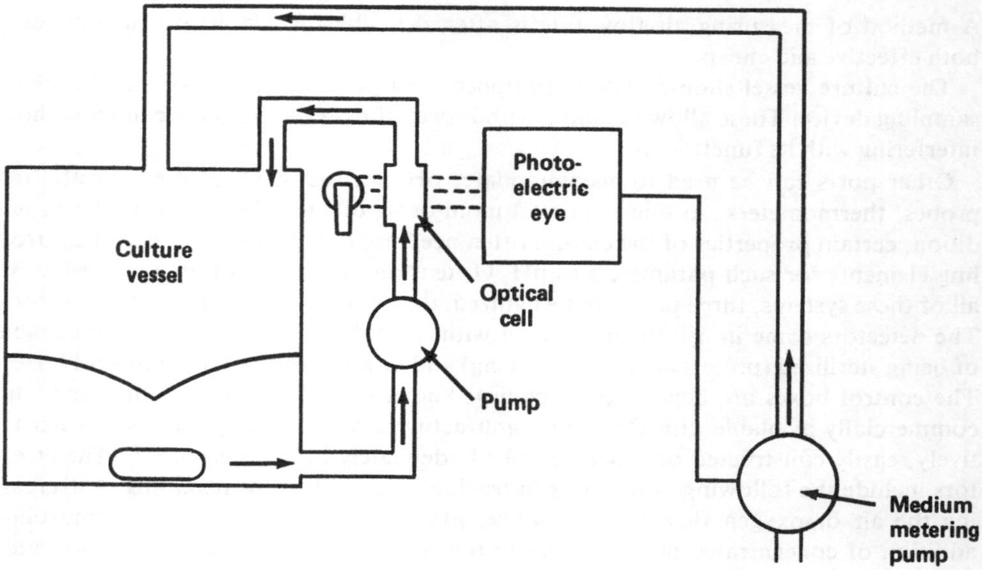

FIGURE 5. Schematic additions necessary to convert chemostat to turbidostat. All materials shown in Figure 2 plus that shown in this figure are required to convert chemostat to turbidostat (for operation see text).

ture is returned to the flask. Many systems have been described.[5-8] Some systems require continuous monitoring, while others are activated for short periods only. Generally, if high cell concentrations are used absorbance is monitored, while with low densities, light scattering is used as a measure. This measurement is then fed to a "black box" (or controller) which with a solenoid-operated valve or peristaltic pump allows medium to enter the culture.

The two main problems in turbidostat operation appear to be due to bubbles entering the photodetector cell and due to cell wall growth on the optical cell surface. The former can be prevented by stopping aeration at time of culture withdrawal, drawing culture from a "bubble-free" area, or allowing bubbles to coalesce in a vertical tube before passage into the photodetector cell.[5-8] Wall growth in the optical cell can not be prevented. It can be minimized or periodically removed. Methods for removal include washing with 5N NaOH with a distilled water rinse[5] or by a "windshield wiper arrangement".[6,8]

E. Spent Medium and Cells

These are collected in a vessel, generally at room temperature. Since several days or weeks of material accumulate, this material is not used for experimental material unless measures are taken to chill or freeze the effluent. The effluent vessel is generally large with an inlet to receive the spent culture media plus an outlet suitably plugged with cotton to allow excess gases to be vented, but to retain aerosolized cells.

IV. OPERATION

A. Media

Any liquid medium may be employed in a continuous culture. Ordinarily, defined media are preferentially used, though for certain studies complex media can be used. For instance, if one wants to compare the effect of nutrition on physiological response and divorce the effect of an altered growth rate, both complex and defined media can be employed with a constant growth rate in a chemostat. Needless to say, in the defined media any nutrient can be used to limit the growth of the organism in chemostat, such as carbon-energy sources, nitrogen source, vitamins, or cofactors. In addition, ion concentration can be adjusted so that they limit growth. These which have been routinely used include potassium, magnesium, phosphate, and iron. Frequently batch cultures are used to ascertain the concentration of the nutrients that will limit the yield of the organisms.

The criteria used to define nutrient limitations are

1. A small increase in concentration of the nutrient in the inflow medium should result in a proportional increase in biomass
2. A substantial alteration in the concentration of other nutrients in the inflow medium should not affect the biomass
3. At low growth rates, the limiting nutrient should not be detectable in the culture vessel, indicating "complete" exhaustion of that nutrient

Postgate and colleagues[17] have described an unusual nutrient limitation of *Azotobacter* by gaseous dinitrogen. Their definition of nutrient limitation does not obey all the above definitions. By stopping the medium flow they observed an increase in biomass indicating that the cells' metabolism of nitrogen was somehow limited. Criterion 2, however, was satisfied.

Chemostat cultures using complex media are of course limited by some nutrient in the medium, but it is unknown to the operator. Turbidostats do not require a limited nutrient to be defined or established.

B. Sterilization

Preferentially, the culture vessels plus medium are autoclaved to sterilize. With certain components of the medium other sterilization procedures are sometimes necessary, such as membrane filtration. With certain types of continuous cultures, particularly those using perspex, ethylene oxide is the only viable alternative. Often the continuous culture is split into the three basic parts, autoclaved separately, and assembled after sterilization. This has the advantage that new media reservoirs and collection reservoirs can be changed when required without disturbing the culture vessel.

C. Inoculation

Once the apparatus has been set up and is found to be working satisfactorily, and before inoculation, the author routinely fills the culture vessel with a small fraction of growth medium, starts the stirrer and air pump. The apparatus is then left 1 day to observe if any organism grows. This is a precaution against the inadvertent contamination of the culture during assembly. Once the sterility of the culture is established, a small sample (5 mℓ) of an overnight culture of the organism to be cultured is added aseptically, through the inoculation port. The organism is grown up in the same medium as that to be used in the continuous culture. Immediately, the medium pump is activated at the desired flow rate. By the time the culture volume is established at the desired value, the cells have usually become established in the culture. An alternative is to fill the culture vessel to the required volume and then inoculate. The culture is allowed to grow to stationary phase before the medium pump is started.

D. Sampling

Routinely, a steady state is established before samples are removed. It is generally accepted that steady states are not attained before about five replacements of the culture have been experienced. The actual time depends on the dilution rate, e.g., at a $D = 0.5\ hr^{-1}$, the time is $5 \times 1/0.5$ hr = 10 hr, while at $D = 0.1\ hr,^{-1}$ the time is $5 \times 1/0.1$ hr = 50 hr. At slow dilution rate the time taken to attain steady-state is progressively longer.

Sampling is accomplished by the sampling device. Small samples (<10% of the culture volume) do not interfere with the steady state of the culture. On the other hand, large samples frequently needed for biochemical assay result in severe depletion of the culture. This does interfere with the steady state. After one of these "bleedings", it is necessary to wait the prescribed time for steady state to be reestablished. The author does not recommend the use of cells from the effluent vessel for study, since these cells have been sitting in a nonphysiologically controlled environment. Certain others advocate the use of these cells;[7] this to the author nullifies the effort needed to set up a continuous culture. Sometimes such large quantities of cells are needed that sampling of the effluent vessel is necessary. Precautions should be taken then to prevent metabolism of the cells either by chilling the reservoir or freezing the cells and spent media as they emerge.

V. SPECIALIZED SITUATIONS

A. Anaerobic Cultures

Not only have aerobes been grown in continuous cultures, but also attempts at growing anaerobes have been successful. Precautions are needed, however. One of the main problems is the diffusion of oxygen through the large surface area of silicone rubber tubing used. This can be minimized by using either PVC or Neoprene tubing, which is less permeable to gases. In addition, a gentle flow of sterile nitrogen (O_2 free) or argon over the surface of the medium in the reservoir prevents reabsorption of oxygen after autoclaving has driven the gases off.[5]

Vigorous stirring is still needed to mix the culture, while a steady stream of sterile argon or nitrogen enables the removal of gaseous products of metabolism, e.g., CO_2. Frequently, nontoxic reductants such as ascorbate or thioglycollate may be added to the medium.

B. High Pressure Systems

Most continuous cultures operate at perhaps 1 to 2 psi above atmospheric pressure. Their construction would not withstand higher pressures. Dalton and Postgate[4] constructed one of perspex capable of withstanding 60 psi to enable the study of effect of increased partial pressure of oxygen and nitrogen.

C. Slow Growing Cultures

Postgate and Hunter,[9] Tempest et al.[10] and Postgate[11] were able to grow *Klebsiella aerogenes* at low growth rates down to the order of $D = 0.004\ hr.^{-1}$ At these growth rates, slight variations in the inflow rate can result in significant changes in the growth rate of the culture. Consequently either very reliable pumps must be utilized, or else culture volume increased, so that higher (and more reproducible) flow rates can be maintained.

D. Pathogenic Organisms in Continuous Culture

Often studies of pathogenic organisms need continuous cultures to be set up. Since these cultures are active for long periods, not only is contamination from the outside increased, but the chance of escape of organisms from culture is increased. The latter is perhaps the most important — at least from safety considerations. Therefore, a net negative pressure within the chemostat to prevent escape of cells is matched by an increased chance of contamination of cells within the apparatus.

Several systems are available, but perhaps the best described is the Porton Mobile Enclosed Chemostat (The POMEC) of Harris-Smith and Evans.[12] In this apparatus, fifteen safety criteria are met. It consists of a culture apparatus in a glove box with various entry and exit ports, process, transfer, and instrument chambers.

E. Industrial and Complex Continuous Cultures

These highly specialized modifications of the basic continuous culture will be discussed in other chapters, principally by Srinivasan, by Drozd and Linton, and by Ricica.

VI. COMPLICATIONS IN THE USE OF CONTINUOUS CULTURE

There are several major complications in the use and operation of continuous cultures. Perhaps the major objection is the complicated nature of the apparatus when compared with the batch cuture, frequently an Erlenmeyer flask with cotton plug. Once the operator has gotten over the reluctance of doing more work than he feels is necessary, the advantages outweigh the disadvantages. However, other complications do arise.

A. Contamination from Outside

Since continuous cultures operate for time periods several orders of magnitude longer than conventional batch cultures, the chance of a contaminant getting into the culture and taking it over is greatly increased. However, with careful precautions, efficient filters on gas outlets and inlets, a functioning autoclave, and a little luck, contamination is really no problem. However, another type of "contamination" is perhaps more important (see next section).

B. Contamination from Inside

Continuous cultures select organisms that grow faster at low substrate concentrations, (i.e., have low K_M for substrate) and continually lose biomass. Consequently, they are fiercely selective. Since growth in continuous culture is maintained for long periods, the chance of a mutation arising is increased. Should a mutation occur to a "more adapted form", this cell would be selected for. Thus it is possible that a mutant could take over the population. In a like manner, pure cultures often have tendencies to form "variants" on culture and this can be significant in continuous cultures. Often these variants "take over" continuous culture and exhibit very different properties to the parent's. The concept and properties of genetic selection will be dealt with in a later chapter by Calcott.

There appears to be two ways of overcoming the problem. One involves maintaining continuous cultures for very short periods before variants take over and never using a continuous culture to inoculate another. This is favored by Tempest.[13] An alternative is advocated by Postgate and Hunter.[9] They believe that continuous cultures should be maintained until the most adapted establish themselves and then experiments started. The problem with this viewpoint is that one never knows when the "final, most successful" variant has established itself. In either case, routine plating of the culture on at least two different types of media allows the operator to become familiar with the colonial morphology of the parent, mutants, and variants that may become established.

C. Foaming

Foaming in a continuous culture is a phenomenon that should be avoided. Not only does it increase the chances of "inoculating" the medium inflow line with cells, but it can interfere with steady states in several ways. By introducing another phase into the culture, the volume of the liquid phase can be affected, resulting in an actual dilution rate different from that determined experimentally. In addition, due to the nature of bacterial cells, they can accumulate in the foam to a cell concentration per mℓ of condensed foam much higher than the culture as a whole. This results in a non-homogenous state in the culture,[15] which obviously can nullify the measurements taken of whole culture properties, e.g., O_2 tension, pH, etc. Foaming appers to be due to fierce aeration of cultures rich in organic materials and those of high cell concentrations.

To reduce or eliminate foaming, the operator can either alter the complexity of the medium, add a low concentration of sodium chloride (1 to 2%), decrease the concentration of cells in the culture by decreasing the concentration of growth-limiting nutrient, or adding antifoam. The first method is obviously not recommended since the medium was chosen for a reason. The second, although effective, can result in an altered cell physiology (see chapter by Ellwood and Robinson). The fourth is perhaps the most used by workers. The antifoam should be capable of being sterilized and be nontoxic and nonmetabolizable. Unfortunately, it can interfere with certain experimental procedures.[13] In addition, often in very heavily foaming cultures, such large quantities of antifoam are needed that they interfere with steady states. In this case, the most effective method is to decrease the cell concentration in the culture vessel.

D. Wall Growth

Wall growth is the growth of cells on the surface of the culture vessel and accessories. With the introduction of a static phase, the homogeneity of the system is destroyed. With heavy wall growth, it is possible to increase the flow rate above the Dc or μ_{max} of the culture, since it is impossible to wash out the culture. In addition, above the water line the wall growth will evaporate leaving a crust that (with time) will fall off, giving a nonhomogenous culture — or operationally worse the crust can get stuck in lines causing disastrous floods as cultures fill up, but do not empty (observation that the author hopes will never be experienced). Consequently, wall growth should be avoided if possible, unless it is required. It has been used by Munson[8] as a selection pressure. Certain cultural conditions appear to promote wall growth, namely certain nutrient limitations, e.g., NH^{+}_{4} — limited growth of *Klebsiella*[16] where large quantities of carbohydrate are accumulated, and when antifoam is used. These conditions can be either avoided or cultures can be maintained only for short periods. If this is impossible, it can be reduced by routinely "scrubbing" the walls of the culture vessel with the stirring bar using a strong external magnet. It is advisable to shut the stirring motor off during this operation for fear the magnet may "bump" across the culture putting delicate apparatus in peril.[5]

A technical type of wall growth on optical faces associated with turbidostats has been dealt with in a previous section.

E. Failure of the Inoculum to Grow

Frequently, after cells are inoculated into the culture vessel they fail to grow — or are very slow in establishing themselves. Routinely, it is advisable to use actively growing cells in the same medium as that used in the continuous culture. Alternatively, it is sometimes effective to "spike" the medium with 0.5% nutrient broth to establish the culture. As the flow is started this spent nutrient broth is washed out. After 5 replacements less than 0.1% of the initial population and broth is present.[5] The cells are usually well established by then.

F. Growth of Cells in the Medium Inlet Line

Cells in continuous culture are aerosolized in the space above the culture. Consequently, there is a real possibility of inoculation of the feed line with cells. Once inoculation and then establishment has been reached, cells can adhere to the wall and grow back even to the medium reservoir. Needless to say, this interferes with the steady state in the culture vessels.

To minimize this, most cultures are set up with one air flow past the medium inlet. However, should growth start and establish itself, there are several methods to deal with it. If stainless steel connectors have been put in the line as "break points", the contaminated part can be removed and replaced by a sterile new part. Alternatively, if growth is limited to just the inlet tube and confined to the first few centimeters, then it is possible to "sterilize" this with an infrared lamp. By heating the glass inlet for 1 hr or so, it is possible to dry out and prevent backgrowth.

G. Mechanical and Electrical Failure

If a mechanical or electrical failure occurs, usually the culture is not lost, though the steady state is upset and a new one must be attained after an appropriate waiting period. A problem rests with automatic pH controllers that on resumption of electrical power may dump several hundreds of milliliters of base or acid into the culture. However, a time delay in the set up can be used.[5] One of the major mechanical problems rests with janitors or cleaners who cannot resist poking continuous cultures. These cultures are extremely sensitive and dislike interference. This can be alleviated by a simple education program.

VIII. COMMERCIALLY AVAILABLE APPARATUS

The researcher wishing to use a continuous culture can either buy a commercially made apparatus or make his own. This section does not intend to cover all the possible suppliers, but will concentrate only on those very familiar to the author (this is a reflection of the activeness of advertising and not the energy of the author). Chemostats are available commercially from several sources:

- New Brunswick Scientific Co. Inc., 1130 Somerset Street, New Brunswick, New Jersey 08903
- Bellco Glass Inc., 340 Edrudo Road, Vineland, New Jersey 08360
- Kontes Glass Co., Vineland, New Jersey 08360

Turbidostats are available commercially from:

- Lab-Line Instruments, Inc., Lab-Line Plaza, 15th and Bloomingdale Avenue, Melrose Park, Illinois 60160
- L. Eschweiler and Co., Kiel, Germany.

Components to make continuous cultures can be purchased from many sources too numerous to mention. The reader is directed to several reviews[2,5,7] and the manufacturer and distributor catalogs in his area.

REFERENCES

1. **Novick, A. and Szilard, L.**, Description of the chemostat, *Science*, 112, 715, 1950.
2. **Baker, K.**, Low cost continuous culture apparatus, *Lab. Pract.*, 17, 817, 1968.
3. New Brunswick Scientific Co., Inc., Catalog, 174, 1978.
4. **Dalton, M. and Postgate, J. R.**, Growth and physiology of *Azotobacter chroococcum* in continuous culture, *J. Gen. Microbiol.*, 56, 307, 1969.
5. **Evans, C. G. T., Herbert, D., and Tempest, D. W.**, The Continuous Cultivation of Micro-organisms, in *Methods in Microbiology*, Vol. 2, Norris, J. R. and Ribbons, D. W., Eds., Academic Press, London, 1970, 278.
6. **Watson, T. G.**, The present status and future perspectives of the turbidostat, *J. Appl. Chem. Biotechnol.*, 22, 229, 1972.
7. **Kubitschek, H. E.**, Introduction to Research with Continuous Cultures, *Prentice-Hall Biological Techniques Series*, Prentice-Hall, Englewood Cliffs, N. J., 1970.
8. **Munson, R. J.**, Turbodistats, in *Methods of Microbiology*, Vol. 2, Norris, J. R. and Ribbons, D. W., Eds., Academic Press, London 1970, 349.
9. **Postgate, J. R. and Hunter, J. R.**, The survival of starved bacteria, *J. Gen. Microbiol.*, 29, 233, 1962.
10. **Tempest, D. W., Herbert, D., and Phipps, P. J.**, Studies on the growth of *Aerobacter aerogenes* at low dilution rates in a chemostat, in *Microbial Physiology and Continuous Culture*, Powell, E. O., Ed., Her Majesty's Stationery Office, London, 1967.
11. **Postgate, J. R.**, The viability of very slow growing populations: a model for the natural ecosystem, *Bull. Ecol. Res. Commun.*, (Stockholm), 17, 287, 1973.
12. **Harris-Smith, R. and Evans, C. G. T.**, The Porton Mobile Enclosed Chemostat (The POMEC) in *Continuous Culture of Microorganisms*, Malek, I., Beran, K., Fencl, Z., Munk, V., Ricica, J., and Smrikova, H., Eds., Academic Press, London, 1969.
13. **Tempest, D. W.**, The place of continuous culture in microbiological research, *Adv. Microb. Physiol.*, 4, 223, 1970.
14. **Lee, S. K. and Calcott, P. H.**, unpublished results.
15. **Goulet, J.**, personal communication.
16. **Calcott, P. H.**, unpublished observation.
17. **Dalton, H. and Postgate, J. R.**, Growth and physiology of *Azotobacter* chroococium in continuous culture, *J. Gen. Microbiol.*, 56, 307, 1969.

Chapter 3

THEORY OF CONTINUOUS CULTURE*

M. J. Bazin

TABLE OF CONTENTS

I. Introduction 28

II. The Equations of Balance 28

III. Substrate-Limited Growth 30
- A. Steady State 30
- B. Monod Growth Kinetics 30
  - 1. Chemostat Culture 31
  - 2. Turbidostat Culture 32
  - 3. Fed-Batch Culture 34
- C. Chemostat Dynamics Close to Equilibrium 35
- D. Transient Behavior of Chemostats 39
- E. Extension of the Monod Model 42
  - 1. Time Delays 42
  - 2. The Contois Model 43
- F. Structured Models 44

IV. Extensions of Basic Chemostat Theory 46
- A. Varying External Environment: Temperature 46
- B. Varying Internal Environment: Chemical Inhibition 47
- C. Multiple Substrate Kinetics 52
- D. Multi-Species Systems 53
- E. Heterogeneous Conditions 55
  - 1. Wall Growth 55
  - 2. Packed Columns 56

References 60

* This chapter was submitted in May 1978.

I. INTRODUCTION

It was not until the fundamental studies of Monod[1] and Novick and Szilard,[2] who formulated the basic theory of the chemostat, that continuous culture techniques began to play a significant role in microbiological research. There now exists a considerable body of literature describing the theory of continuous culture, and several excellent accounts of the subject exist.[3-6] In this chapter I have not tried to write an exhaustive review, but have attempted to illustrate, by specific examples, some of the methods that have been used to define, describe, and analyze continuous culture systems. I have placed most emphasis on chemostat culture, because it is by far the most commonly employed method of continuous culture.

The chapter is organized along the following lines: in Section II the physical basis and underlying assumptions of a substrate-limited continuous culture is described in theoretical terms. From this description chemostat, fed-batch, and turbidostat cultures are defined. The next section describes the behavior of a chemostat at and close to steady state. Most emphasis is given to models based on Monod[7] substrate-limited growth kinetics, but examples of other formulations are given also. Ways in which some of the underlying assumptions of chemostat culture may be relaxed are described next. The effect of temperature, chemical inhibition, and the theory of multisubstrate and multispecies systems is dealt with briefly. Finally, the effect of wall growth and the theory of continuous flow columns is introduced.

II. THE EQUATIONS OF BALANCE

Consider the idealized continuous flow system depicted in Figure 1. Here we have a vessel containing biomass at density x in a volume of culture V. Fresh medium is supplied and culture is removed continuously at flow rate, F. We make the following assumptions:

1. All the organisms are distributed randomly in the culture vessel, i.e., the cells do not adhere to each other or to the walls of the vessel, and the suspension is well mixed.
2. The population density, x, is a continuous variable, i.e., although the population consists of a number of discrete particles, their concentration is sufficiently high and the size of each particle sufficiently small for this discreteness to be ignored, and the volume occupied by the particles compared to the total volume of the system is negligible. Thus biomass is regarded as being dissolved in the culture solution.
3. As soon as nutrient enters the culture vessel it is instantenously dispersed throughout the culture.
4. No mutation of the organism occurs.

If we let the specific growth rate of the organism be μ, then the change in population density within the culture vessel is

$$\frac{dx}{dt} = \mu x - \frac{F}{V} x \qquad (1)$$

We specify that:

5. The specific growth rate is a function of some single, growth-limiting nutrient, i.e., $\mu = \mu(s)$, where s is the concentration of the nutrient in the culture vessel.

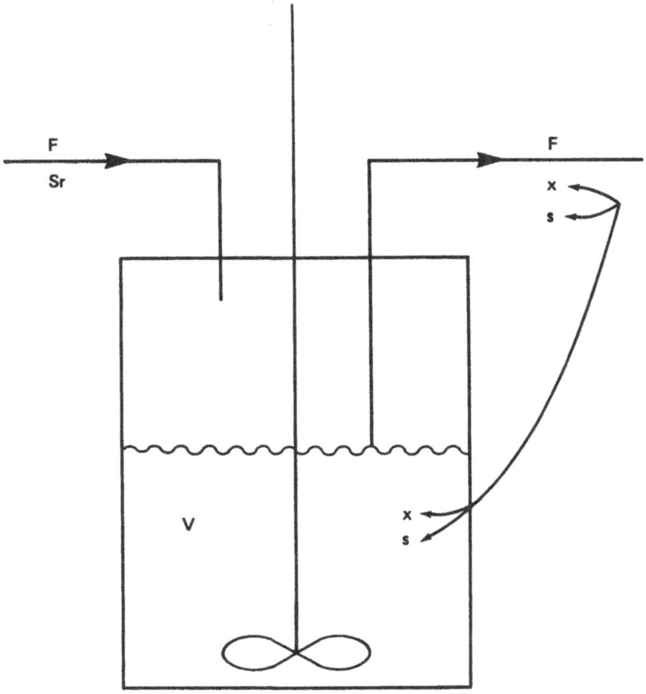

FIGURE 1. Idealized continuous flow systems consisting of a well-mixed culture of volume V fed with nutrient at rate F and concentration S_r. Biomass at density X and unused substrate at concentration s is removed at the same rate.

And make the following further assumptions:

6. A constant amount of biomass is produced per unit of growth-limiting nutrient utilized. This constant is the growth yield, Y.
7. Growth inhibition by product formation does not occur.
8. The external environment of the culture vessel remains constant.
9. All organisms are viable.

Based on these assumptions mass balance equations for both biomass and limiting nutrient may be written:

$$\frac{dx}{dt} = \mu(s)x - \frac{F}{V}x \tag{2}$$

$$\frac{ds}{dt} = \frac{F}{V}S_r - \frac{\mu(s)x}{Y} - \frac{F}{V}s \tag{3}$$

where S_r is the concentration of growth-limiting nutrient supplied to the culture.

Equations 2 and 3 may be regarded as the fundamental equations of balance for a substrate-limiting continuous culture system. The behavior of the dependent variables x and s with respect to time reflects the dynamics of the system, which in turn depends on the form of $\mu(s)$, the function specifying the kinetics of substrate-limited growth. We may distinguish immediately three types of continuous culture — depending on which quantity in Equations 2 and 3 is chosen as a parameter. When the volume of the culture remains constant, and the flow rate controlled so that the dilution rate, D = F/V, is a parameter, the system becomes a chemostat. By photomechanical mechanisms the biomass density may be controlled — in which case the system becomes a turbidostat. A fed-batch culture may be defined as one in which the volume term in Equations 2 and 3 is a function (usually linear) of time, i.e., V = V(t).

Equations 2 and 3 are usually written in a less general form to accomplish one or more of the following:

1. Test hypotheses associated with the kinetics of microbial growth usually, although not invariably, by substituting specific functions for $\mu(s)$.
2. Reduce the number of assumptions — or modify them — so that they correspond more closely to experimental feasibility.
3. Extend the physical basis of the system by considering feedback loops, reactors in series, etc.
4. Extend the biological aspects of the system by considering more than one specific and/or limiting nutrient.

Herbert[3] and Pirt[4] have described in detail theoretical aspects of feedback in continuous cultures and chemostats in series, and the reader is referred to these authors for further information about item 3 above. Proceeding sections of this chapter contain examples of the other three ways in which Equations 2 and 3 have been extended to investigate theoretically the way in which microorganisms grow.

III. SUBSTRATE-LIMITED GROWTH

A. Steady State

When a specific function for $\mu(s)$ is substituted into Equations 2 and 3, the resulting set of differential equations are invariably nonlinear, and analytical solutions for them do not exist. A variety of mathematical or computer-based techniques may be applied to approximate a solution, but most emphasis has been placed on analysis after the system has reached steady state. This may be defined as the condition arising when all the extensive variables of the system are constant with respect to time, i.e.,

$$\frac{dx}{dt} = \frac{ds}{dt} = 0 \qquad (4)$$

Under such conditions, the mathematics of the system usually become tractable and the behavior of the dependent variables at steady state, designated $\tilde{x}$ and $\tilde{s}$, as a function of some parameter such as D or S_r can be calculated. One of the most important properties of chemostat culture is realized immediately when the steady-state condition is imposed on Equation 1. By setting the derivative equal to zero, the specific growth rate becomes equal to the dilution rate (D = F/V) of the system. Thus by using chemostat culture the experimentalist has the capacity to precisely control the specific growth rate of a microbial population by adjusting the rate of nutrient supply to the culture vessel and allowing the system to come to steady state.

B. Monod Growth Kinetics

The most widely employed substrate-dependent specific growth rate function is that attributed to Monod[7] and takes the following form:

$$\mu(s) = \frac{\mu_m s}{K_s + s} \qquad (5)$$

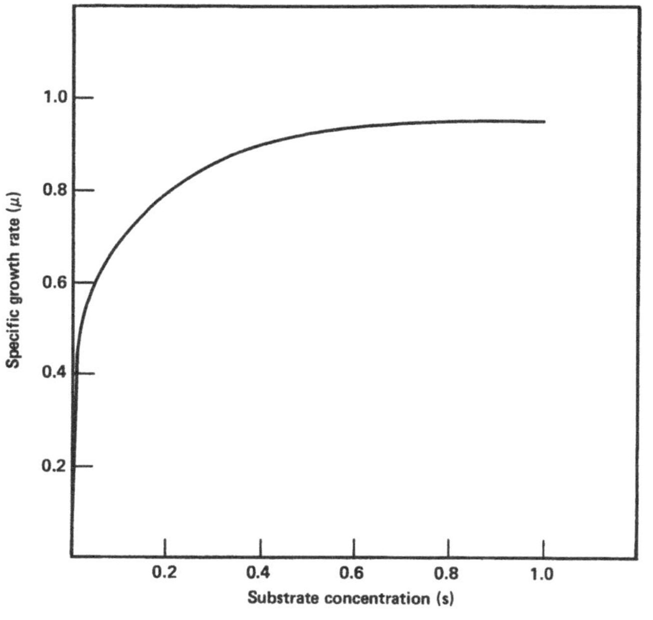

FIGURE 2. Relationship between the specific growth rate of a microbial population and the concentration of a limiting nutrient according to the relationship of Monod given in Equation 5 with μ_m = 1 and K_s = 0.05.

This relationship is shown in Figure 2, where μ_m can be seen to be the maximum specific growth rate attainable — which is approached asymptotically as increases and K_s, the saturation constant, is the concentration of limiting nutrient required for growth at half the maximum rate.

Substituting Equation 5 into Equations 2 and 3 gives

$$\frac{dx}{dt} = \frac{\mu_m xs}{K_s + s} - Dx \tag{6}$$

$$\frac{ds}{dt} = D(S_r - s) - \frac{\mu_m xs}{Y(K_s + s)} \tag{7}$$

It is sometimes supposed that the Monod function can be derived along analogous lines to the Michaelis-Menton expression for enzyme kinetics, but such derivations are applicable strictly only to steady-state conditions.[8] When steady state is assumed, the resulting algebraic equations derived from Equations 6 and 7 can be rearranged to give the following relationships for the three types of continuous culture that have been mentioned.

1. Chemostat Culture

$$\tilde{s} = \frac{DK_s}{\mu_m - D}, \quad \mu_m > D \tag{8}$$

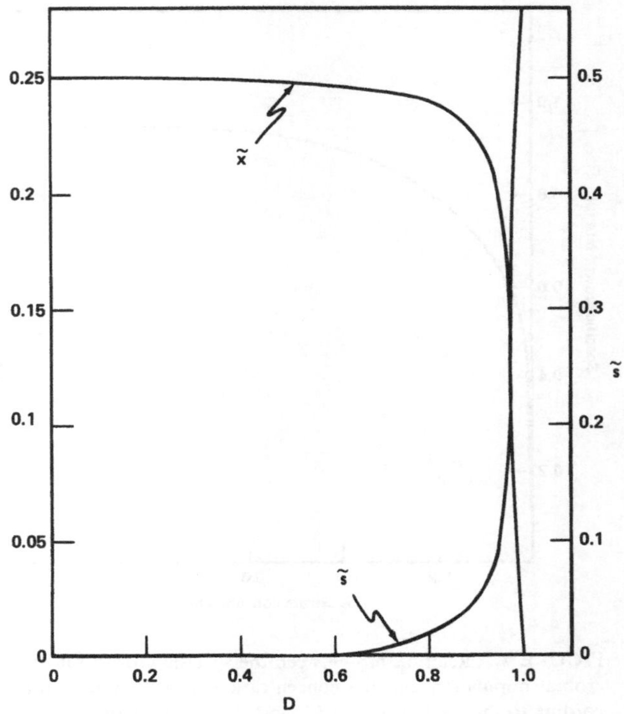

FIGURE 3. Steady-state nutrient ($\tilde{S}$) and biomass concentration ($\tilde{X}$) as a function of dilution rate in a chemostat culture, based on Monod[7] kinetics with Y = 0.5, μ_m = 1.0, K_s = 0.005, and S_r = 0.5.

$$\tilde{x} = Y(S_r - \tilde{s}), \; S_r \geq \frac{DK_s}{\mu_m - D} \tag{9}$$

Otherwise

$$\tilde{s} = S_r$$
$$\tilde{x} = 0$$

Steady-state biomass and substrate concentrations are plotted against dilution rate and input nutrient concentration in Figures 3 and 4. For S $\geq DK_s/(\mu_m - D)$, S is independent of S_r, and x is directly proportional to S_r. At lower values of S_r, washout of biomass is expected. The critical dilution rate, D_c, is defined as the value of D when s is set equal to S_r in Equation 8. At dilution rates above D_c the culture washes out.

2. *Turbidostat Culture*

$$\tilde{s} = S_r - \frac{\tilde{x}}{Y} \tag{10}$$

$$D = \frac{\mu_m \tilde{s}}{K_s + \tilde{s}} \tag{11}$$

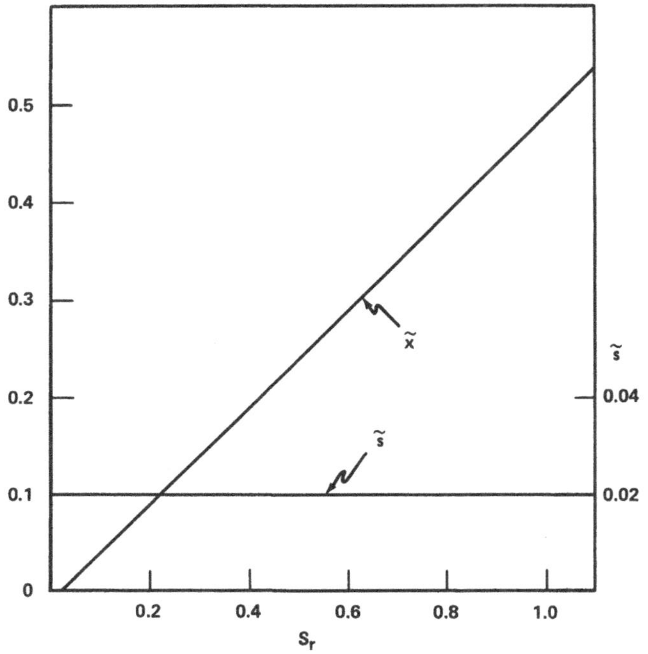

FIGURE 4. Biomass density and limiting nutrient as a function of input nutrient concentration at steady-state in a chemostat. Symbols and growth constants as in Figure 3. Dilution rate = 0.8.

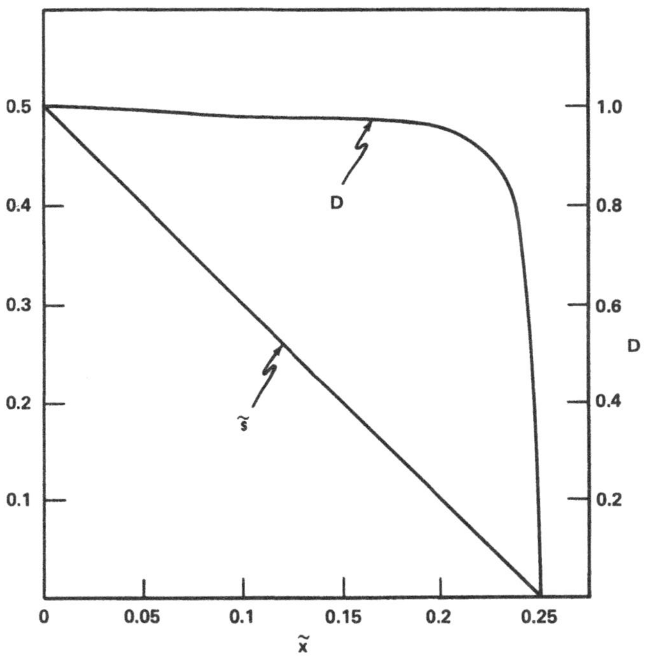

FIGURE 5. Steady-state limiting nutrient concentration at dilution rate as a function of biomass density in turbidostat culture. Symbols and growth constants as in Figure 3.

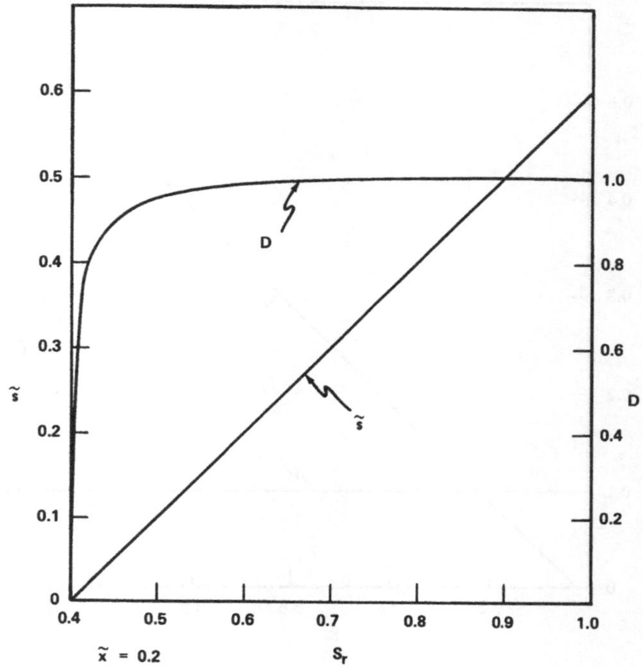

FIGURE 6. Relationship between dilution rate, substrate concentration, and input nutrient concentration at steady-state in a turbidostat culture with constant $\tilde{X}$ = 0.2. Symbols and growth constants as in Figure 3.

Here $\tilde{s}$ is linearly dependent on both $\tilde{x}$ and S_r, while D is a hyperbolic function of these two parameters. These relationships are illustrated in Figures 5 and 6.

3. Fed-Batch Culture

In this case, we make the assumption that although D = F/V(t) the change in dilution rate is so slow that steady-state or "quasi-steady-state"[4] conditions can occur. When V(t) is a linear function,

$$V(t) = V_o + Ft \tag{12}$$

where V_o is the culture volume at time t = 0, then

$$D = \frac{F}{V_o + Ft} \tag{13}$$

$$\tilde{s} = \frac{DK_s}{\mu_m - D}, \quad \mu_m > D \tag{14}$$

$$\tilde{x} = Y(S_r - \tilde{s}), S_r \geqslant \frac{DK_s}{\mu_m - D} \tag{15}$$

As shown in Figure 7, the dilution rate decreases with time, but its relative rate of change is slow compared to that of the biomass and substrate concentrations as the

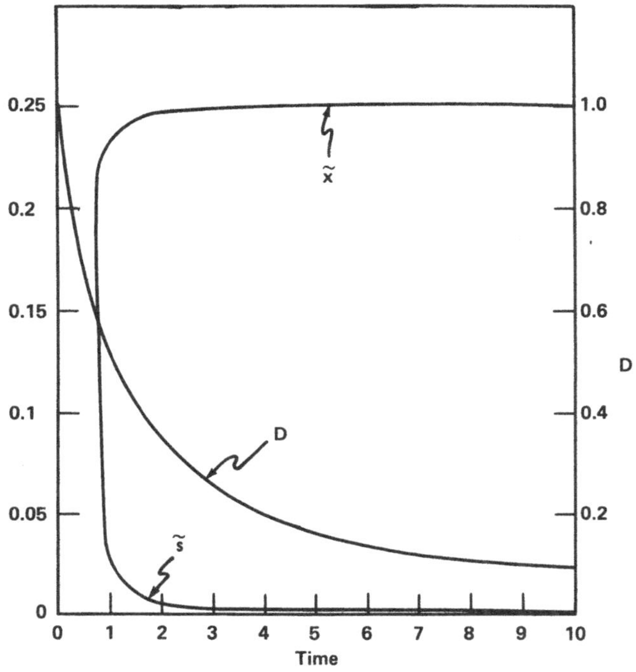

FIGURE 7. Dilution rate, limiting nutrient concentration, and biomass density as a function of time in a fed-batch culture assuming "quasi-steady-state".[4] Symbols as in Figure 3, $S_r = 0.5$, $K_s = 0.005$, $\mu_m = 0.6$, $Y = 0.5$, $V_o = 10.0$, $F = .0$.

system rapidly reaches a situation of maximal substrate utilization. During the early stages of fed-batch culture, however, the dilution rate may be higher than μ_m, in which case a steady state cannot be attained.

C. Chemostat Dynamics Close to Equilibrium

After specifying the function $\mu(s)$ in Equations 2 and 3, and providing appropriate estimates of the parameters and constants involved, the resulting nonlinear model may be simulated either on an analog computer or numerically on a digital computer. An introduction to simulation techniques applied to biological systems is given by Patten,[9] and many texts on this topic exist. Such methods are limited in their lack of generality, however, and it is often advantageous to postpone their application until it is necessary to compare the prediction of a dynamic model to experimental results. A better understanding of dynamic systems may be obtained initially by more analytical methods. An example of such an approach is the method of Liapounov, which provides information about the stability of dynamic systems close to equilibrium. Here the method will be applied to a simple chemostat system based on Monod kinetics and in later sections it will be used in describing growth inhibition by a microbial product and multispecies systems.

For a chemostat, Equations 2 and 3 may be rewritten as

$$\frac{dx}{dt} = \mu(s)x - Dx \qquad (16)$$

$$\frac{ds}{dt} = Ds_r - \frac{\mu(s)x}{Y} - Ds \qquad (17)$$

Equilibrium or stationary points are defined by the values of x and s when the system is at steady state, i.e., $\tilde{x}$ and $\tilde{s}$. In terms of the Monod function, Equations 16 and 17 have only one nontrivial equilibrium point given by Equations 8 and 9. In applying the Liapounov method, we seek to determine the behavior of the system close to the equilibrium point. To do this we define new variables, X and S, which have their equilibrium at the origin,

$$X = x - \tilde{x}$$
$$S = s - \tilde{s} \qquad (18)$$

Substituting into Equations 16 and 17 gives

$$\frac{dX}{dt} = \mu(S + \tilde{s})(X + \tilde{x}) - D(X + \tilde{x}) \qquad (19)$$

$$\frac{dS}{dt} = DS_r - \frac{\mu(S + \tilde{s})(X + \tilde{x})}{Y} - D(S + \tilde{s}) \qquad (20)$$

We now expand μ in a Taylor series and drop second order terms:

$$\mu(S + \tilde{s}) = \mu(\tilde{s}) + S\mu'(\tilde{s}) \qquad (21)$$

Here the prime rotation represents a derivative. By specifying the Monod function,

$$\mu(\tilde{s}) = \frac{\mu_m \tilde{s}}{K_s + \tilde{s}} \qquad (22)$$

and

$$\mu'(\tilde{s}) = \frac{(K_s + \tilde{s})\mu_m - \mu_m \tilde{s}}{(K_s + \tilde{s})^2} \qquad (23)$$

Substituting into Equation 19 yields

$$\frac{dX}{dt} = \left[\frac{\mu_m \tilde{s}}{K_s + \tilde{s}} + \frac{(K_s + \tilde{s})\mu_m - \mu_m \tilde{s}}{(K_s + \tilde{s})^2} S\right][X + \tilde{x}] - D(X + \tilde{x})$$

$$= \frac{\mu_m \tilde{s} X}{K_s + \tilde{s}} + \frac{\mu_m \tilde{s}\tilde{x}}{K_s + \tilde{s}} + \frac{(K_s + \tilde{s})\mu_m XS}{(K_s + \tilde{s})^2} + \frac{(K_s + \tilde{s})\mu_m S\tilde{x}}{(K_x + \tilde{s})^2}$$

$$- \frac{\mu_m \tilde{s} SX}{(K_s + \tilde{s})^2} - \frac{\mu_m S\tilde{s}\tilde{x}}{(K_s + \tilde{s})^2} - DX - D\tilde{x} \qquad (24)$$

Here the second term and the last term define the steady-state equations for dx/dt = 0 and therefore disappear. As we are considering the behavior of X and Y close to equilibrium only, and at their equilibrium they both have a value of zero, all we are concerned with are small values of X and Y. Second order terms in these variables will have very small values that are approximate to zero. Therefore the third and fifth terms in Equation 24 can be eliminated. After some cancellation and rearrangement this leaves

$$\frac{dX}{dt} = \left[\frac{\mu_m \tilde{s}}{K_s + \tilde{s}} - D\right] X + \left[\frac{\mu_m \tilde{x}}{K_s + \tilde{s}} - \frac{\mu_m \tilde{s}\tilde{x}}{(K_s + \tilde{s})^2}\right] S \quad (25)$$

Replacing $\tilde{S}$ and $\tilde{X}$ by the relationships in Equations 8 and 9 gives

$$\frac{dX}{dt} = a_{11} X - a_{12} S \quad (26)$$

where

$$a_{11} = 0$$

and

$$a_{12} = \frac{[\mu_m - D]\, Y\, [Sr(\mu_m - D) - K_s D]}{K_s \mu_m}$$

By going through a similar procedure with Equation 20, we obtain

$$\frac{dS}{dt} = a_{21} X - a_{22} S \quad (27)$$

where

$$a_{21} = -D/Y$$

and

$$a_{22} = -(D + a_{12}/Y)$$

The two linearized Equations 26 and 27 are now used to obtain an analytical solution for the system in terms of X and S, and therefore x and s close to equilibrium. On differentiation Equation 26 gives

$$\frac{d^2X}{dt^2} = a_{11}\frac{dX}{dt} - a_{12}\frac{dS}{dt}$$

$$= a_{11}\frac{dX}{dt} - a_{12}(a_{21}X - a_{22}S)$$

$$= a_{11}\frac{dX}{dt} - a_{12}a_{21}X - a_{22}\left(\frac{dX}{dt} - a_{11}X\right) \quad (28)$$

On rearranging we find

$$\frac{d^2X}{dt^2} - (a_{11} + a_{22})\frac{dX}{dt} - (a_{12}a_{21} - a_{11}a_{22})X = 0 \quad (29)$$

which has the solution

$$X = A_1 e^{\lambda_1 t} + A_2 e^{\lambda_2 t} \tag{30}$$

where A_1 and A_2 are constants associated with the initial conditions of the system and the eigenvalues λ_1 and λ_2, are given by the solution of the characteristic equation,

$$\lambda^2 - (a_{11} + a_{22})\lambda - (a_{12}a_{21} - a_{11}a_{22}) = 0 \tag{31}$$

which is solved by applying the quadratic formula:

$$\lambda = \frac{[a_{11} + a_{22}] \pm [(a_{11} + a_{22})^2 + 4(a_{12}a_{21} - a_{11}a_{22})]^{1/2}}{2} \tag{32}$$

Solving for λ and simplifying we find

$$\lambda_1 = -a_{12}/Y$$

and

$$\lambda_2 = -D \tag{33}$$

With the steady-state condition that $D = \mu_m\tilde{s}/(k_s + \tilde{s})$, both λ_1 and λ_2 are negative real numbers. Thus X in Equation 30 will tend to zero smoothly and monotonically as time, t, becomes large so that x will tend to $\tilde{x}$ in the same way. This type of solution is called a stable node, and suggests that the biomass density moves towards its equilibrium point without oscillating. Analyzing S shows that substrate concentration behaves in a similar manner.

The analysis described above indicates that the qualitative behavior of a linearized dynamic system is determined by the nature of its eigenvalues. In the example of a simple chemostat system, both of these were negative real numbers and so the system was found to be stable. If either or both had been positive, the solution of Equation 30 would be unstable, as in such circumstances X increases without limit. When only one eigenvalue is positive the solution is called a saddle point, and when both are positive it is an unstable node. Another type of solution occurs when the expression within the square root sign in Equation 32 is negative. In such cases the eigenvalues are in the form of conjugate pairs,

$$\lambda = n \pm pi \tag{34}$$

where n and p are real numbers, and i is defined as the square root of −1. With such eigenvalues the solution of Equation 30 is sinusoidal. For $n < 0$ the oscillations in X are damped and the system moves towards steady state. Such a solution is called a stable focus. If n is positive, oscillations of increasing amplitude occur, an unstable focus. When the size of such fluctuations is limited by some internal constraint such as the amount of nutrient provided, undamped oscillations occur, the amplitudes of which are independent of the initial conditions of the system. Such behavior constitutes

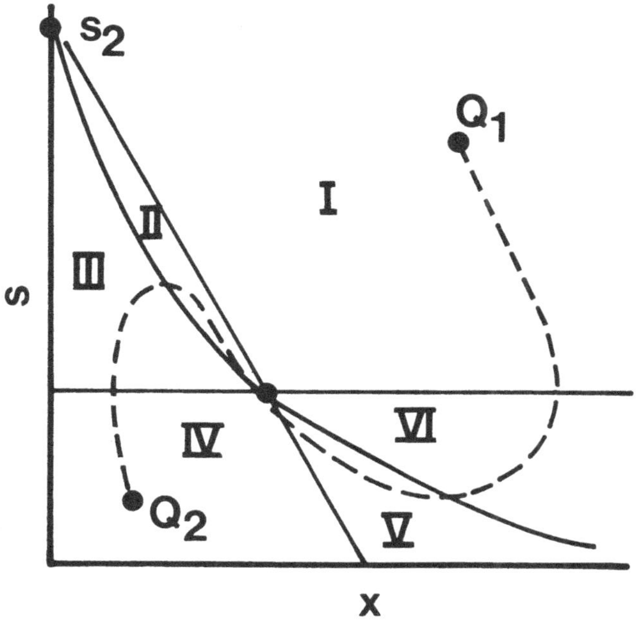

FIGURE 8. Phase plane plot of limiting nutrient concentration against biomass density from which the transient behavior of a chemostat can be estimated. Roman numerals represent subdomains of the positive quadrant delineated by the separatrics represented by the solid lines and defined by Equations 38 to 40. (After Koga, S. and Humphrey, A. E., *Biotechnol. Bioeng.*, 9, 375, 1967. With permission.)

a limit cycle. Finally, a vortex is obtained when n is zero. In this case sustained oscillations occur also, but their amplitude depends on the initial conditions.

Quite often it is convenient to write linearized differential equations in matrix form, particularly when more than two variables are involved, and determine eigenvalues from the associated determinant. When equations 26 and 27 are written this way they become

$$\begin{bmatrix} \dot{X} \\ \dot{S} \end{bmatrix} = \begin{bmatrix} a_{11} & a_{12} \\ a_{21} & a_{22} \end{bmatrix} \begin{bmatrix} X \\ S \end{bmatrix} \qquad (35)$$

where the dot over a variable represents its derivative with respect to time. The characteristic equation defining the eigenvalues is

$$\begin{vmatrix} a_{11} - \lambda & a_{12} \\ a_{21} & a_{22} - \lambda \end{vmatrix} = 0 \qquad (36)$$

D. Transient Behavior of Chemostats

Although the eigenvalues in Equation 35 indicate that the variables in a Monod-based chemostat will never oscillate, Koga and Humphrey[10] have shown that the system

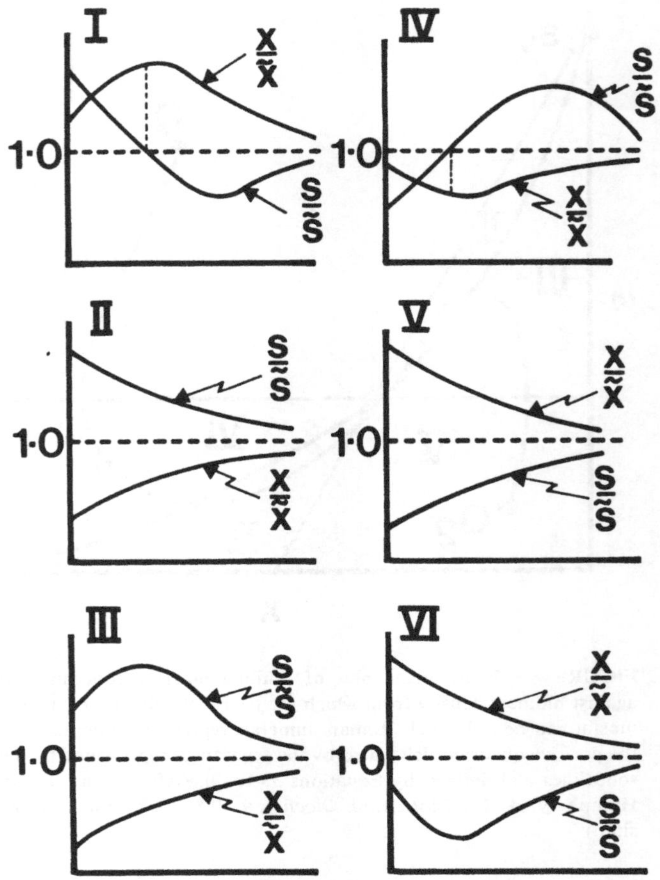

FIGURE 9. Behavior of chemostat cultures with initial conditions represented by each of the subdomain in the phase plane plot of Figure 8. (After Koga, S. and Humphrey, A. E., *Biotechnol. Bioeng.*, 9, 375, 1967. With permission.)

can theoretically overshoot or undershoot in its transient response. Dividing Equation 6 by Equation 7 gives

$$\frac{dx}{ds} = \frac{\left(\frac{\mu_m s}{K_s + s} - D\right) x}{D(S_r - s) - \frac{\mu_m s x}{Y(k_s + s)}} \tag{37}$$

By determining the sign of the derivative, dx/ds, in Equation 37, Koga and Humphrey[10] drew the phase plane shown in Figure 8. The domain in the positive quadrant is subdivided into six subdomains by three separatrics defined by

$$s = \frac{DK_s}{\mu_m - D} \tag{38}$$

$$s = S_r - \frac{x}{Y} \tag{39}$$

$$D(S_r - s) = \frac{\mu_m sx}{Y(K_s + s)} \qquad (40)$$

Consider the point Q in Figure 8. It lies within subdomain I so that

$$x > Y(S_r - s) \qquad (41)$$

from Equation 39, and

$$s > \frac{DK_s}{\mu_m - D} \qquad (42)$$

implied by Equation 38. Applying these relationships to Equation 37 reveals that the denominator must be negative, while the numerator must be positive. Thus $dx/ds < 0$. In subdomain VI,

$$x > Y(S_r - s) \qquad (43)$$

and

$$s > \frac{DK_s}{\mu_m - D} \qquad (44)$$

This makes both the denominator and the numerator in Equation 39 negative, so that $dx/ds > 0$. By analysis of this sort the dashed line in Figure 8 can be generated. Figure 9 shows the response of the system for a trajectory starting in each subdomain. Both undershoots and overshoots in x and s can occur as they approach steady state.

Perram[11] has analyzed the chemostat equations based on Monod kinetics in terms of how long it takes for the system to come to steady-state. By adding suitable multiples of Equations 6 and 7 we obtain

$$\frac{d}{dt}\left(\frac{x}{Y} + s\right) = DS_r - D\left(\frac{x}{Y} + s\right) \qquad (45)$$

which on integration gives

$$\frac{x}{Y} + s = S_r - \left[S_r - \left(\frac{x_o}{Y} + s_o\right)\right] e^{-Dt} \qquad (46)$$

where x_o and s_o represent the concentrations of biomass and limiting nutrient in the culture vessel at time zero. In the limit, as t approaches infinity,

$$\frac{x}{Y} + s = S_r \qquad (47)$$

That is to say the system comes to steady state. The quantity 1/D, the reciprocal of term multiplying t in the exponent in Equation 46, Perram[11] calls the relaxation time. Characteristically it takes the system two or three relaxation times to come to steady state. Perram[11] found that his theoretical estimates of the time required for a chemostat to come to steady state were much shorter than those found experimentally. He concluded, therefore, that a second, much longer relaxation time was hidden in the descriptive differential equations. This second relaxation time was found by considering the system close to steady state so that Equation 47 can be used to eliminate x in Equation 7. This gives

$$\frac{ds}{dt} = \frac{(\mu_m - D)(S_r - s)(\theta - s)}{K_s + s}$$

where

$$\theta = \frac{DK_s}{\mu_m - D} \qquad (48)$$

On integration,

$$c\,\frac{(\Theta - s)^{K_s + \Theta}}{(S_r - s)^{K_s + S_r}} = \exp\left[-(S_r - \Theta)(\mu_m - D)t\right] \qquad (49)$$

where c is a constant of integration that cannot be found by solving for the initial conditions at t = 0, because Equation 48 is valid only for the steady state when $t \ll 1/D$. Rearranging the exponent on the right hand side of Equation 49 gives

$$(S_r - \Theta)(\mu_m - D) = \mu_m S_r - D(S_r + K_s) \qquad (50)$$

Perram[11] derives two relaxation times depending on whether the right hand side of Equation 50 is positive or negative. With $\mu_m s_r > D(k_s + S_r)$ the relaxation time is given by

$$\frac{S_r + K_s}{D(S_r + K_s) - \mu_m S_r}$$

which corresponds to washout of the culture. When $\mu_m s_r < D(k_s + S_r)$ the relaxation time becomes

$$\frac{K_s \mu_m}{[\mu_m - D]\,[\mu_m S_r - D(K_s + S_r)]}$$

At values of D close to μ_m the time for a chemostat culture to come to steady state may be long which, as Perram[11] points out, might explain experimental observations.

E. Extension of the Monod Model

Although the Monod equation is the most widely quoted model of substrate-dependent microbial growth, several modifications to it have been suggested. Two of these are described in this section and other examples are introduced subsequently.

1. Time Delays

According to the Monod expression a change in limiting nutrient concentration results in an immediate change in the specific growth rate of the microbial populations.

This is an unlikely eventuality, as it must take a finite amount of time for more nutrient to be processed or for the organisms to adapt to lower concentrations of nutrient. One of the simplest ways to take this lag in response time into account is to include a time delay term into the Monod function. Caperon[12] used a discrete time delay, but probably a more realistic approach is that of MacDonald[13] who used a continuous term replacing s in Equation 5 by

$$s = \int_0^\infty s(t - \tau)F(\tau)d\tau \qquad (51)$$

Two different functional forms for $F(\tau)$ were studied by introducing a new variable, z, to replace s in the differential equation describing the change in biomass in a chemostat, and adding a third equation for dz/dt. With

$$F(\tau) = a \exp(-a\tau), \qquad (52)$$

which lets the most recent values of s have the greatest effect on z, the chemostat equations become

$$\frac{dx}{dt} = \frac{\mu_m xz}{K_s + z} - Dx \qquad (53)$$

$$\frac{ds}{dt} = D(S_r - s) - \frac{\mu_m xz}{Y(K_s + z)} \qquad (54)$$

$$\frac{dz}{dt} = a(s - z) \qquad (55)$$

The coefficient a in these equations corresponds to the scale of the time delay with a long delay being produced by small values of a. Linear stability analysis of Equations 53 to 55 reveals that damped oscillations can occur when a is small. Sustained limit cycle oscillations were obtained with

$$F(\tau) = a^2\tau \exp(-a\tau) \qquad (56)$$

in which case Equation 55 was replaced by

$$\frac{dz}{dt} = a(w - z) \qquad (57)$$

and a fourth differential equation was introduced to define the new variable w

$$\frac{dw}{dt} = a(s - w) \qquad (58)$$

2. The Contois Model

The two coefficients in the Monod model, μ_m and K_s, have been given functional dependence on a variety of environmental factors such as temperature, pH, and inhibitor concentration. Some of the ways in which this has been accomplished will be mentioned later. Contois,[14] on the basis of experimental evidence from chemostat cultures of *Aerobacter aerogenes,* proposed a model in which the saturation constant is a function of biomass density so that the specific growth rate function becomes

$$\mu(s) = \frac{\mu_m s}{K_s x + s} \tag{59}$$

Using this function in the chemostat equations and solving for steady-state conditions yields the following relationships:

$$D = \frac{\mu_m \tilde{s}}{K_s \tilde{x} + \tilde{s}} \tag{60}$$

$$\tilde{x} = Y(S_r - \tilde{s}) \tag{61}$$

Rearrangement of these equations shows that there is a linear relationship between the reciprocal of the dilution rate and the reciprocal of the steady-state substrate concentration:

$$\frac{1}{D} = \frac{YK_sS_r}{\mu_m} \cdot \frac{1}{\tilde{s}} + \left(\frac{1}{\mu_m} - \frac{YK_s}{\mu_m}\right) \tag{62}$$

The equivalent relationship when the Monod function is used is

$$\frac{1}{D} = \frac{K_s}{\mu_m} \cdot \frac{1}{\tilde{s}} + \frac{1}{\mu_m} \tag{63}$$

In the latter case, the slope of the straight line generated by plotting $1/D$ against $1/\tilde{s}$ is independent of the input nutrient concentrations. Contois[14] found experimentally that this was not the case, but that the slope depended on S_r as predicted by Equation 62.

F. Structured Models

The inclusion of time delays or density-dependent terms into models of microbial growth may go some way towards predicting the behavior of continuous cultures, but it does not add much to our knowledge of the relationship between microbial population dynamics and the growth and division of cells. Models that reflect the physiological state of the cell population have been called structured models.[15] A simple and early example of such a model is that of Williams,[16] who considered growth of a microbial population in a chemostat to follow the pattern illustrated in Figure 10. The model supposes that biomass density can be divided into two parts, synthetic (c) and structural/genetic (n). Nutrient is absorbed from the culture medium into the synthetic portion and is then converted into structural/genetic material. Biomass density (m) is the sum of c and n. Reaction rates are assumed to be proportional to the quantities present, and bimolecular. The dynamics of the system are described by

$$\frac{ds}{dt} = D(S_r - s) - k_1 sm \tag{64}$$

$$\frac{dc}{dt} = k_1 sm - k_2 cn - Dc \tag{65}$$

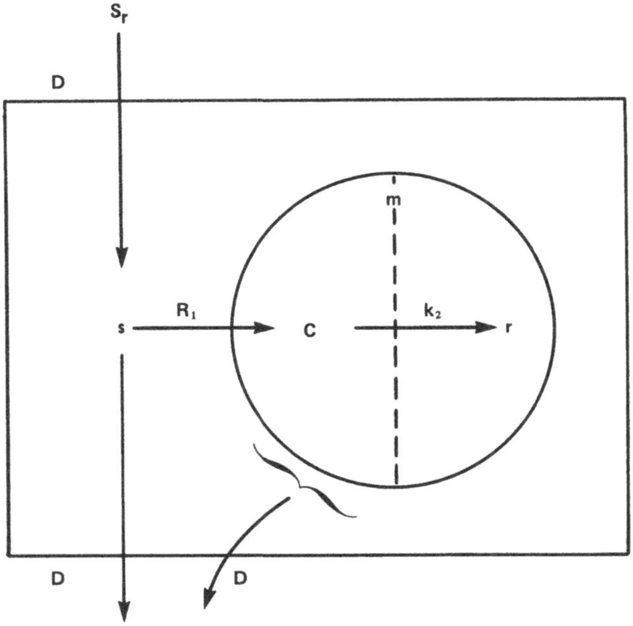

FIGURE 10. The structured model of Williams[16] in which biomass density is composed of a c-portion, which is converted to the n-portion. Nutrient at concentration s in a chemostat is supplied at concentration S_r. Nutrient is supplied and removed at dilution rate D.

$$\frac{dn}{dt} = k_2cn - Dn \tag{66}$$

$$m = n + c \tag{67}$$

Williams[16] equates the n-portion of his model with cell number density. In contrast to simpler, unstructured models this allows the behavior of both extensive variables (such as biomass density) and intensive variables (such as mean cell mass (m/n)) to be described. Steady-state predictions of Williams' model are shown in Figure 11, and Figure 12 illustrates the predicted behavior of a chemostat starting with a stationary phase inoculum. These results bear both qualitative and quantitative similarities to the behavior of real systems. Even closer agreement has been obtained by further modifications of the model,[17,18] but obtaining a good fit to data is not entirely the goal of Williams' modeling exercise. Probably the most significant aspect of his results is that by imposing a simple (and quite possibly unrealistic) structure on the way biomass is represented, fairly complex and realistic behavior can be simulated. As Williams[16] points out, the approach he employs shows the extent to which oversimplified presuppositions can give rise to biologically interesting results.

Other structural growth models have been proposed.[15,19,20] According to Fredrickson,[21] most of them seem to have one feature in common — that they are incorrectly formulated! Fredrickson[21] examined the models of Williams[16] and several other workers, and found that none of them contained terms for the dilution of intracellular components. For example, in Williams'[16] model the c and n portions will be diluted by expansion (growth) of the total biomass, but this is not taken into account in the equations of balance.

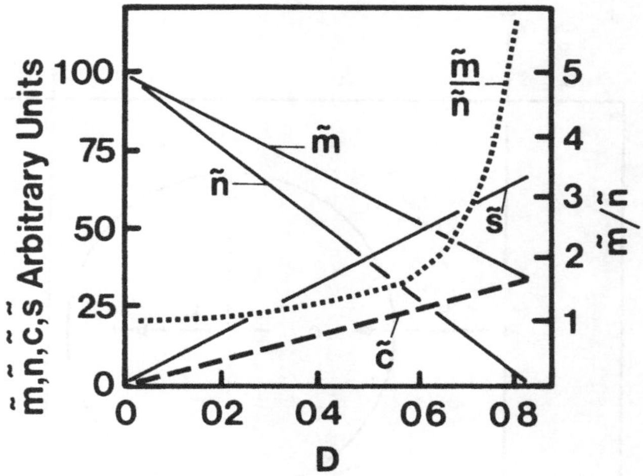

FIGURE 11. Steady-state predition of Williams[16] structured model given by Equations 64 to 67. (After Williams, F. M., *J. Theor. Biol.*, 15, 190, 1967. With permission.)

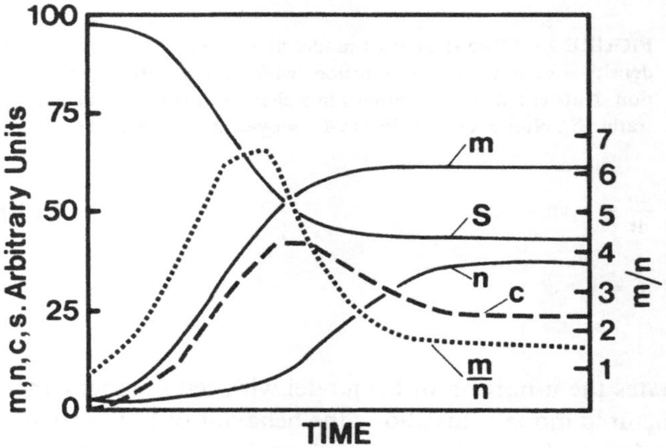

FIGURE 12. Behavior of a chemostat inoculated with stationary phase cells according to the structured model of Williams.[16] (After Williams, F. M., *J. Theor. Biol.*, 15, 190, 1967. With permission .)

IV. EXTENSIONS OF BASIC CHEMOSTAT THEORY

In this section we present examples of how some of the basic assumptions of an ideal continuous culture system mentioned in Section II may be relaxed or changed.

A. Varying External Environment: Temperature

As an example of how the assumption of constant external environment can be dropped, we will discuss some of the ways chemostat theory has been used to investigate the effect of temperature on microbial growth kinetics. The relationship between temperature and specific growth rate in terms of the Monod function was investigated by Knowles et al.[22] They found that for the nitrifying bacteria, *Nitrosomonas* and *Nitrobacter*, both μ_m and K_s were exponential functions of temperature (T) between 8°C and 23°C, i.e.,

$$\mu_m(T) = Ae^{aT}$$
$$K_s(T) = Be^{bT} \qquad (68)$$

where A, B, a, and b are constants.

Topiwala and Sinclair[23] extended this work by showing that between 25°C and 40°C K_s followed Arrhenius-type kinetics in *Klebsiella aerogenes:*

$$\frac{1}{K_s(T)} = A \exp(-aRT) \qquad (69)$$

where R is the gas constant. Between 25°C and 45°C they fitted μ_m to a double reciprocal relationship:

$$\mu_m(T) = B \exp(-b/RT) - C \exp(-c/RT) \qquad (70)$$

These relationships and values for the appropriate constants were determined from steady-state chemostat cultures. Topiwala and Sinclair[23] then simulated the effect of temperature shifts at constant dilution rate and compared their predictions to experimental results. They found that a lag between stimulus and response occurred that was not predicted by their theory. A response lag was incorporated into their model by replacing T with what they called "effective temperature" (T′) such that

$$T' = T + J\left[1 - \exp(-t/\tau)\right] \qquad (71)$$

Here J is the value of the step-change in temperature and can take on positive or negative values depending on whether a step-up or step-down shift was employed. Although τ represents a time constant in this relationship, Topiwala and Sinclair[23] found that it was dependent on the magnitude and direction of the step-change. Typical results obtained by these authors are shown in Figure 13.

Williams[16] discussed the effect of temperature on a less empirical basis by incorporating temperature-dependent terms into his structural model described in the last section. By letting the two rate constants, k_1 and k_2, become functions of temperature he derived an expression relating the size of organisms growing in a chemostat to the temperature of the culture.

B. Varying Internal Environment: Chemical Inhibition

Several kinetic models, mostly based on modifications of the Monod equation, have been proposed to describe the kinetics of microbial growth inhibition. Edwards[24] analyzed a variety of substrate inhibition models, the simplest of which is

$$\mu(s) = \frac{\mu_m K_i s}{K_s K_i + K_i s + s^2} \qquad (72)$$

where K_i is an "inhibitor" constant. When this function is substituted into the chemostat equations and the system solved for steady-state, we find

$$\tilde{s}^2 + \left(K_i - \frac{\mu_m}{D}\right)\tilde{s} + K_s K_i = 0 \qquad (73)$$

and

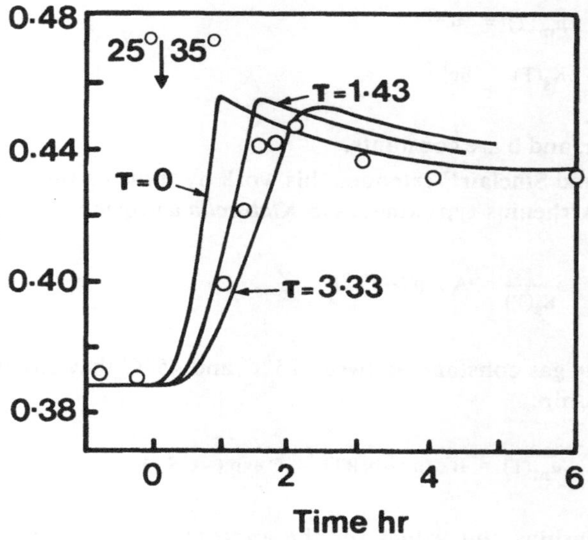

FIGURE 13. Transient response of a chemostat after a shift-up in temperature according to the model of Topiwala and Sinclair.[23] Solid lines represent predictions at three values of τ (see text) and open circles are data. (After Topiwala, H. and Sinclair, C. G., *Biotechnol. Bioeng.*, 13, 795, 1971. With permission.)

$$\tilde{x} = Y(S_r - \tilde{s}) \tag{74}$$

The quadratic form of Equation 73 implies that for a given dilution rate two steady states are possible, as illustrated in Figure 14. The nature of these steady states can be ascertained by Liapounov stability analysis. Linearization of the chemostat equations containing $\mu(s)$ defined by Equation 72 gives

$$\begin{bmatrix} \dot{X} \\ \dot{S} \end{bmatrix} = \begin{bmatrix} a_{11} & a_{12} \\ a_{21} & a_{22} \end{bmatrix} \begin{bmatrix} X \\ S \end{bmatrix} \tag{75}$$

where X and S are defined in Equation 18 and:

$$a_{11} = o$$

$$a_{12} = \frac{\tilde{x}}{\tilde{s}} \left[1 - \frac{D}{\mu_m} \left(1 + \frac{2\tilde{s}}{K_i} \right) \right]$$

$$a_{21} = -D/Y$$

$$a_{22} = -\left[D + \frac{a_{12}}{Y} \right]$$

The eigenvalues, found by solving the characteristic Equation 36, are

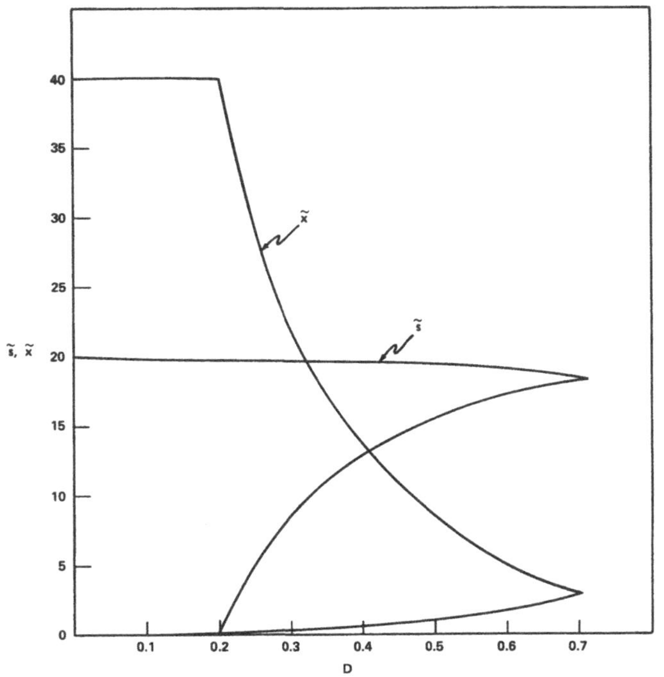

FIGURE 14. Relationship between limiting nutrient and dilution rate at steady-state in a chemostat subject to the substrate limitation function of Equation 72 with $S_r = 40.0$, $K_s = 1.0$, $K_i = 10.0$, $\mu_m = 1.0$, and $Y = 0.5$.

$$\lambda_1 = -\frac{a_{12}}{Y} \qquad \lambda_2 = -D \tag{76}$$

Both eigenvalues are real and one of them, λ_2, is always negative. The stability of the system therefore depends solely on λ_1. If

$$\frac{\mu_m}{D} > 1 + \frac{2\tilde{s}}{K_i} \tag{77}$$

then a_{12} is positive and λ_1 is negative In this case the steady state is stable. On the other hand if

$$\frac{\mu_m}{D} < 1 + \frac{2\tilde{s}}{K_i} \tag{78}$$

then λ_1 is positive and the solution is a saddle point and the system is unstable. The inequalities in Equations 77 and 78 can be used to compare the stability of two steady states occurring at one dilution rate. Let the steady-state concentration of substrate at the maximum value of μ be $\tilde{s}_m$ and let $\tilde{s}_1$ and $\tilde{s}_2$ be two steady-state substrate concentrations at a given value of D such that

$$\tilde{s}_1 \leq \tilde{s}_m \leq \tilde{s}_2 \tag{79}$$

and

$$\mu(\tilde{s}_1) = \mu(\tilde{s}_2) = D \tag{80}$$

We find $\tilde{s}_m$ by equating to zero the differential of $\mu(\tilde{s})$ with respect to $\tilde{s}$:

$$\frac{d\mu(\tilde{s})}{d\tilde{s}} = 0 \Longrightarrow \frac{K_s}{\tilde{s}^2} = \frac{\tilde{s}}{K_i} \tag{81}$$

Therefore

$$\frac{K_s}{\tilde{s}_1^2} \geqslant \frac{\tilde{s}_1}{K_i} \tag{82}$$

and

$$\frac{K_s}{\tilde{s}_2^2} \leqslant \frac{\tilde{s}_2}{K_i} \tag{83}$$

Rearrangement of Equation 80 gives

$$\frac{\mu_m}{D} = \frac{K_s}{\tilde{s}_1} + 1 + \frac{\tilde{s}_1}{K_i} = \frac{K_s}{\tilde{s}_2} + 1 + \frac{\tilde{s}_2}{K_i} \tag{84}$$

so that

$$\frac{\mu_m}{D} \geqslant 1 + \frac{2\tilde{s}_1}{K_i} \tag{85}$$

and

$$\frac{\mu_m}{D} \leqslant 1 + \frac{2\tilde{s}_2}{K_i} \tag{86}$$

Applying these results to the relationships in 77 and 78 we see that $\tilde{s}_1$ implies a stable steady state while $\tilde{s}_2$ is unstable. As $\tilde{s}_1$ is smaller than $\tilde{s}_2$, this means that the stable situation is one in which more biomass is produced.

The analysis above is based on the work of Yano and Koga[25] who considered a more generalized model of substrate inhibition. These authors found no sustained oscillation in the system. This contrasts to results in some of their later work on the theory of inhibition by a microbial product in chemostat culture.[26] Pirt[4] derives expression for both competitive and noncompetitive inhibition. Yano and Koga[26] studied the latter using a general equation, the simplest form of which is

$$\mu(s) = \frac{\mu_m s}{K_s + \left(1 + \frac{\eta}{K_\eta}\right) s} \tag{87}$$

Here K_η is an inhibition constant and η the concentration of an inhibitory product.

When this expression is incorporated into the chemostat equations, an extra differential equation is required to describe the change in η. Yano and Koga[26] suggest

$$\frac{d\eta}{dt} = \left(\eta_1 - \frac{\eta_2 \mu_m s}{K_s + \left(1 + \frac{\eta}{K_\eta}\right) s}\right) x - D\eta \quad (88)$$

Here η_1 and η_2 indicate the way in which product is supposed to be produced and consumed. The following possibilities may be distinguished:

- Completely growth associated: $\eta_1 = 0$; $\eta_2 > 0$
- Completely nongrowth associated: $\eta_1 > 0$; $\eta_2 = 0$
- Partially growth associated: $\eta_1 > 0$; $\eta_2 > 0$
- Negatively growth associated: $\eta_1 > 0$; $\eta_2 < 0$

Analysis near equilibrium produces a set of three linked linear equations. Koga and Yano[26] simplify the system by supposing that growth-limiting substrate is supplied in excess so that at any given moment, $s \gg K_s$. A change in s will therefore have little effect on $\dot{x}$ and $\dot{p}$ so that the system can be reduced to two variables and two equations:

$$\frac{dx}{dt} = \frac{\mu_m x}{1 + \frac{\eta}{K_\eta}} - Dx \quad (89)$$

$$\frac{d\eta}{dt} = \frac{(\eta_1 + \eta_2)\mu_m x}{1 + \frac{\eta}{K_\eta}} - D\eta \quad (90)$$

Linearizing near the single positive equilibrium point of these equations yields the following eigenvalues:

$$\lambda = -\frac{D}{2}\left\{1 + \frac{\left(1 - \frac{D}{\mu_m}\right)}{\left(1 + \frac{\eta_1}{\eta_2 D}\right)}\right\} \pm \frac{D}{2}\left\{\left[1 + \frac{\left(1 - \frac{D}{\mu_m}\right)}{\left(1 + \frac{\eta_1}{\eta_2 D}\right)}\right]^2 - 4\left[1 - \frac{D}{\mu_m}\right]\right\}^{1/2} \quad (91)$$

where

$$0 < D < \mu_m$$

The behavior of the system depends on the nature of product formation in the following ways:

1. Completely growth-associated system ($\eta_1 = 0$, $\eta_2 > 0$). Both values of λ are real and negative. The solution is therefore a stable node and after a small perturbation it will move smoothly and monotonically back to steady state.
2. Completely nongrowth associated system ($\eta_1 > \eta_2 = 0$). In this case the term $(2 - D/\mu_m)/(1 + \eta_1/\eta_2 D)$ tends to zero as η_2 approaches zero. In the limit, Equation 91 becomes

$$\lambda = -\frac{D}{2} \pm \frac{D}{2}\left(-3 + \frac{4D}{\mu_m}\right)^{1/2} \tag{92}$$

The only positive steady-state solution is stable because both eigenvalues are negative. When $0 < D < 0.75\ \mu_m$, the eigenvalues are complex conjugates, and so the system is a stable focal point and exhibits damped oscillations.

3. Partially growth-associated system ($\eta_1 > 0$, $\eta_2 > 0$). Here either a stable node or a stable focus can ccur.
4. Negatively growth-associated system ($\eta_1 > 0$, $\eta_2 < 0$). Let

$$f(D) = -\frac{D}{2}\left[1 + \frac{\left(1 - \frac{D}{\mu_m}\right)}{\left(1 + \frac{\eta_1}{\eta_2 D}\right)}\right] \tag{93}$$

If $f(D) < 0$, the eigenvalues of Equation 91 are both either real, negative, or complex with negative real parts. Hence a stable steady state is reached with or without oscillations. If $f(D) > 0$, at least one of the two eigenvalues is real, positive, or complex with positive real parts. In either case the solution is unstable, with or without oscillations. In the latter case limit cycles are expected. Finally, if $f(D) = 0$, the solution of Equation 91 is a pure imaginary number and sustained, initial conditions-dependent oscillations of the vortex type occur.

C. Multiple Substrate Kinetics

The Monod theory has been extended to include circumstances where more than one required nutrient is present at limiting concentrations. Two conflicting hypotheses have been proposed to describe the effect on the specific growth rate of a culture under such circumstances. Ryder and Sinclair[27] and Sykes[28] assume that at any given time only one such substrate can effectively limit growth. Thus for two growth-limiting substrates at concentrations s_1 and s_2, with Monod growth constants represented by appropriate subscripts

$$\mu(s) = \frac{\mu_{m_1} s_1}{K_{s_1} + s_1} \tag{94}$$

when

$$\frac{\mu_{m_1} s_1}{K_{s_1} + s_1} < \frac{\mu_{m_2} s_2}{K_{s_2} + s_2}$$

and

$$\mu(s) = \frac{\mu_{m_2} s_2}{K_{s_2} + s_2} \tag{95}$$

when

$$\frac{\mu_{m_1} s_1}{K_{s_1} + s_1} > \frac{\mu_{m_2} s_2}{K_{s_2} + s_2}$$

Megee et al.,[29] on the other hand, propose a double substrate model in which both nutrients affect the specific growth rate simultaneously:

$$\mu(s) = \frac{\mu_{m_1} s_1 s_2}{(K_{s_1} + s_1)(K_{s_2} + s_2)} \tag{96}$$

Which of these models is more appropriate has not been determined, although Bader et al.[30] argue in favor of the double substrate model by showing that unless one of the substrates is in relatively high concentrations, both are likely to affect the specific growth rate at the same time. These authors admit, however, that there is little to distinguish the two models when they are considered in terms of a chemostat culture at steady state.

D. Multi-Species Systems

Several investigators have modeled the behavior of more than one species growing in continuous culture.[31-33] In this section, the interaction between a microbial predator and its prey in a chemostat will be discussed. The elementary theory of microbial predation in continuous culture has been reviewed by Curds and Bazin,[34] and Canale[35] has analyzed such a system, based on Monod kinetics, close to equilibrium. Here predator-prey interaction is used to illustrate a generalized approach to the analysis of chemostat systems that does not involve proposing specific kinetic models of microbial growth.[36]

Consider a chemostat in which the growth of a microbial predator at density q is limited by the prey density, h. Let the prey feed on a limiting nutrient present in the culture vessel at concentration s. The equations of balance for the system are

$$\frac{ds}{dt} = DS_r - Ds - hf(s) \tag{97}$$

$$\frac{dh}{dt} = hf(s) - Dh - qg(h) \tag{98}$$

$$\frac{dq}{dt} = qg(h) - Dq \tag{99}$$

Here we have assumed that growth of the prey depends only on its limiting nutrient concentration, and growth of the predator is a function of prey density alone. For simplicity, yields are considered to be unity. We now make what we suppose to be reasonable assumptions about the functions of f and g. For example, when no nutrient is present f = 0; when no prey is present g = 0 and as s increases so does f. With reasoning like this we choose the following assumptions:

$$f(0) = g(0) = 0$$

$$\frac{df}{ds} > 0, \quad \frac{dg}{dh} > 0$$

$$f(\infty) = F, \quad g(\infty) = G; \quad F, G \text{ finite.}$$

Setting Equations 97 to 99 equal to zero, we find that the system has three stationary points:

1. $\tilde{h} = \tilde{q} = o, \; s = S_r$
2. $\tilde{h} = S_r - f^{-1}(D), \; \tilde{q} = o, \; \tilde{s} = f^{-1}(D)$
3. $\tilde{h} = g^{-1}(D), \; \tilde{q} \neq o, \; \tilde{s} \neq o$

Of these only in 3 do both prey and predator survive.

Applying Liapounov's method we obtain the following characteristic equation defining the three eigenvalues of the system:

$$\begin{vmatrix} -D - h\dfrac{\partial f}{\partial s} - \lambda & -f & o \\ \tilde{h}\dfrac{\partial f}{\partial s} & f - D - \tilde{q}\dfrac{\partial g}{\partial h} - \lambda & -g \\ o & \tilde{q}\dfrac{\partial g}{\partial h} & g - D - \lambda \end{vmatrix} = o \tag{100}$$

As has been stated previously, the stability of a system such as this depends on whether its eigenvalues are positive or negative. It is possible to determine this without actually solving the characteristic equation by applying what are known as the Routh-Hurwitz criteria. By writing a characteristic equation in polynomial form,

$$\lambda^n + a_1\lambda^{n-1} + \ldots\ldots + a_{n-1}\lambda + a_n = o \tag{101}$$

the Routh-Hurwitz criteria for n = 2 are

$$a_1 > o \quad \text{and} \quad a_2 > o,$$

for n = 3,

$$a_j > o \quad (j = 1, 2, 3)$$

and $a_1 a_2 > a_3$

for n = 4,

$$a_j > o \; (j = 1, 2, 3, 4)$$

$$a_1 a_2 > a_3$$

$$a_1 a_2 a_3 > a_1^2 a_4 + a_3^2$$

The formula for the Routh-Hurwitz criteria for polynomials of higher order than four is given in texts on differential equations, such as that of Sanchez.[37]

Applying the Routh-Hurwitz criteria to Equation 100 we find that the only one criterion for stability that is not identically satisfied is

$$f - \tilde{h}\frac{\partial f}{\partial s} - \tilde{q}\frac{\partial g}{\partial h} - D < o \tag{102}$$

so it is on the basis of this inequality that the stability of the system stands, and it sets the conditions for a stable predator-prey system in a well-mixed, open ecosystem such as a chemostat. By inserting specific functions for f and g into Equation 102, it is possible therefore to determine the stability properties of models conforming to the given assumptions. Details of this analysis are given by Saunders and Bazin,[36] who also include the effects of arbitrarily increasing S_r (enrichment) and extending the functional dependence of f and g.

E. Heterogeneous Conditions

The assumption of homogeneity in chemostat cultures is an important one as it allows a wide range of simplifications to be made. In practice, however, many continuous cultures are not homogeneous.

1. Wall Growth

One of the commonest causes of inhomogeneity, even in well-stirred systems, is wall growth. A simple model for this type of inhomogeneity has been proposed by Topiwala and Hamer[38] under the following assumptions:

1. Microorganisms adhere in a monolayer to the walls of the culture vessel in contact with the liquid phase.
2. The organisms adhering to the walls reach a constant density after an initial transient period, and their progeny are released into the culture.
3. The growth rate of the attached microorganisms is identical to that of those suspended in the culture medium.

With the total amount of attached biomass proportional to a constant, A, the chemostat equations become

$$\frac{dx}{dt} = \frac{\mu_m xs}{K_s + s} + \frac{\mu_m As}{K_s + s} - Dx \tag{103}$$

$$\frac{ds}{dt} = D(S_r - s) - \frac{\mu_m xs}{Y(K_s + s)} - \frac{\mu_m As}{Y(K_s + s)} \tag{104}$$

and at steady state

$$(DY - \mu_m Y)\tilde{s}^2 + (\mu_m A + \mu_m YS_r - DYS_r + DYK_s)\tilde{s} - DYS_rK_s = 0 \tag{105}$$

When this equation is solved for positive values of s and steady-state biomass density estimated from Equation 9, the result shown in Figure 15 is obtained. Wall growth is seen to effectively extend the range of dilution rates over which a chemostat can be operated. This implies that productivity (Dx) with respect to dilution rate can be increased by increasing the value of A. Topiwala and Hamer[38] suggest that for optimum biomass production it might be profitable to increase the surface area/volume ratio in order to increase monolayer wall growth in single-stage continuous culture systems.

The theory of wall growth under conditions of substrate inhibition as defined by Equation 72 has been investigated by Howell et al.[39] They allowed the thickness of biomass on the walls of the culture vessel to exceed a monolayer, but considered the total amount of attached biomass to be constant. In contrast to the situation in a well-

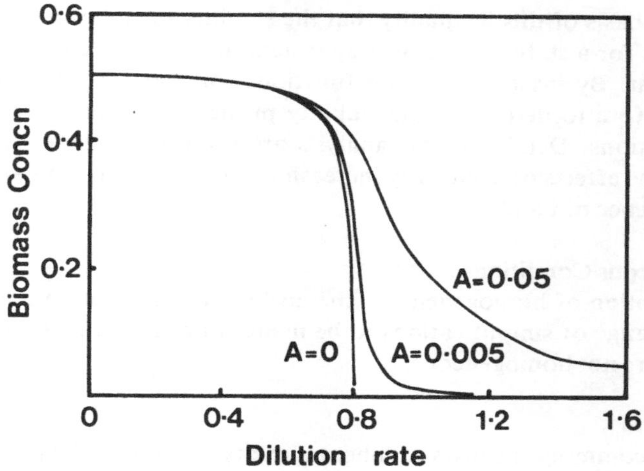

FIGURE 15. Effect of wall growth on steady-state biomass in chemostat culture according to the model of Topiwala and Hamer[38] defined by Equations 103 and 104. With A = 0, no wall growth occurs. (After Topiwala, H. W. and Hamer, G., *Biotechnol. Bioeng.*, 13, 919, 1971. With permission.)

mixed reactor subject to substrate inhibition in which only one stable equilibrium occurs, with wall growth two stable steady states are possible. The conditions necessary for the existence and stability of these multiple steady states are defined by Chi et al.[40] As these authors point out, the assumption of constant biomass density and activity on the walls of the culture vessel might advantageously be relaxed in subsequent analysis of the system. It should also be noted that wall growth quite often occurs to such an extent as to significantly reduce the culture volume. Thus the dilution rate will increase with time as accretion of biomass continues, and the system will take on some of the properties of a fed-batch culture.

2. *Packed Columns*

The analysis of wall growth described above shows that continuous culture dynamics may be significantly affected by biomass adhering to solid surfaces, and that in some circumstance microbial adhesion may be advantageous. One way of increasing microbial attachment is to increase surface area availability by introducing solid particulate material into a culture. Mixing in such a system can be achieved by feeding air and/or nutrient under pressure through the bottom of the culture vessel, in which case the system becomes a fluidized bed reactor. The theory of such cultures is described in detail by Atkinson.[41] Packed columns in which nutrient is fed at the top and allowed to trickle down through the packing material to which the microorganisms are attached have some properties in common with the percolating filter method of effluent purification, slow sand filters, the "quick vinegar" process for acetification of alcohol, and have been used as an experimental model of a soil ecosystem.[42] The chief features distinguishing a packed column culture from a chemostat culture are that microbal growth is assumed to occur on the packing material (glass beads, soil particles, clinker, etc.), the system is not stirred and nutrient (and metabolic product) flow is directional with input entering at the top of the column and effluent emitted from the bottom.

As a result of unidirectional flow in a packed column, concentration gradients of nutrients, products, and biomass occur. The dynamics of a packed column therefore depend not only on time, but upon position in terms of distance down the column.

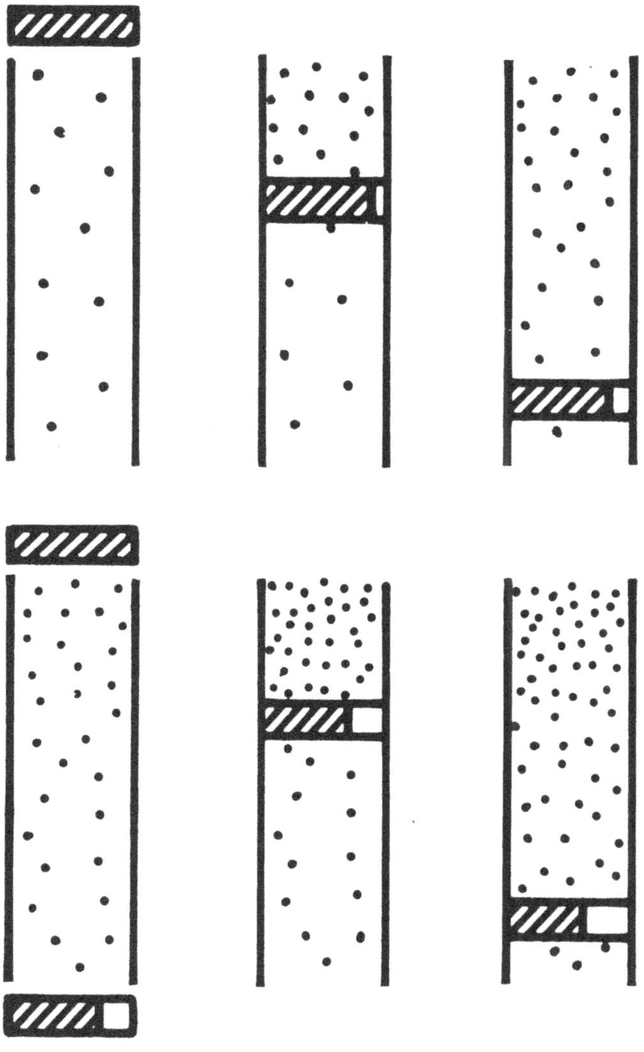

FIGURE 16. Schematic representation of microbial growth on a substrate in a continuous flow column showing how nutrient and biomass density change as a function of both time and distance down the column. Biomass density is represented by the density of the dots and the concentration of nutrient by the length of the shaded portion in the rectangles. (After Saunders, P. T. and Bazin, M. J., *Soil. Biol. Biochem.*, 5, 545, 1973. With permission.)

This property is explained diagrammatically in Figure 16. Equations describing similar systems (such as the decay of a catalyst in a continuous flow reactor) have been derived by chemical engineers. A simple derivation and application to a microbiological system is given by Saunders and Bazin.[43] We consider a column of constant cross-section packed with inert particulate material such as glass beads and containing an initial population of microorganisms, which is uniformly distributed throughout the column. We assume that:

1. Neither dispersion nor hydrodynamic diffusion occur.
2. All of the microorganisms adhere to the solid substratum.

3. Oxygen and other metabolites are in excess.
4. Products are not growth limiting.
5. Product (p) formation is first order with respect to nutrient (s) concentration.
6. The specific rate of reaction is proportional to the biomass density (x) of the microorganisms.

The equations of balance for such a system are

$$\frac{\partial s}{\partial t} + q\frac{\partial s}{\partial u} = -kxs \tag{106}$$

$$\frac{\partial p}{\partial t} + q\frac{\partial p}{\partial u} = kxs \tag{107}$$

$$\frac{\partial x}{\partial t} = f(x, s) \tag{108}$$

Here u is the distance down the column and q is the specific discharge of the system defined as the average net rate of flow of fluid across a unit cross-section of the column.

It is apparent immediately that the mathematical description of this system involves partial differential equations, rather than the set of ordinary differential equations used to describe a well-mixed chemostat. This is a consequence of the dependence of the variables on distance down the column as well as time.

The solutions of Equations 106 to 108 depend, of course, on the nature of the function, f, used to describe the growth of the microbial population. Almost inevitably it becomes necessary to analyze a set of nonlinear, partial differential equations. Broadly speaking two approaches to the dynamic analysis of such sets of equations have been used. One method subdivides the column into theoretical compartments, each of which is considered to be a well-mixed system, so that the fairly straightforward methods of numerical integration used for sets of ordinary differential equations can be employed. The other method seeks to make reasonable simplifications so that the theoretical disruption of the system can be reduced to a form that is mathematically tractable.

A simple application of the compartmentalization approach is given by Prosser and Gray.[44] These authors considered the conversion of nitrite to nitrate by *Nitrobacter* in a column packed with glass beads. Each of their theoretical compartments into which they subdivided the column can be conidered to be equivalent to a chemostat with all the biomass fixed as wall growth, with nutrient and product entering from the compartment above, and effluent going to the compartment below. In general form the equations of balance for the jth compartment ($j > 1$) are

$$\frac{ds_j}{dt} = D(s_{j-1} - s_j) - g(s_j, x_j) \tag{109}$$

$$\frac{dx}{dt} = f(s_j, x_j) \tag{110}$$

$$\frac{dp_j}{dt} = D(p_{j-1} - p_j) + g(s_j, x_j) \tag{111}$$

$$\text{For } j = 1,\ s_{j-1} = S_r;\ p_{j-1} = o$$

Here g is a function describing the conversion of substrate (nitrite, s) to product (nitrate, p). The dilution rate, D, is defined as the rate of flow of material through the column divided by the volume of liquid in the column. Prosser and Gray[44] provided explicit functions for f and g in Equations 109 and 111 and simulated the dynamic behavior of the column by numerically integrating the set of ordinary differential equations in each compartment. In order to determine the transient behavior of the system after a change in flow rate, it was necessary to ensure that the residence time (Θ) remained the same, otherwise the results before and after the shift would not be comparable. This was achieved by making the total number of compartments (n) a function of flow rate (F) in the following way:

$$n = \frac{V}{F\Theta} \tag{112}$$

In this manner a change in flow from 10 cm^3 hr^{-1} to 5 cm^3 hr^{-1} would double the number of theoretical compartments considered in the simulation.

Equations 106 to 108 can be investigated analytically by a simple change of variables:

$$\xi = u \quad \text{and} \quad \tau = t - \frac{u}{q} \tag{113}$$

By the Chain Rule, we know that

$$\frac{\partial s}{\partial u} = \frac{\partial s}{\partial \xi}\frac{\partial \xi}{\partial s} + \frac{\partial s}{\partial \tau}\frac{\partial \tau}{\partial u} \tag{114}$$

and

$$\frac{\partial s}{\partial t} = \frac{\partial s}{\partial \xi}\frac{\partial \xi}{\partial t} + \frac{\partial s}{\partial \tau}\frac{\partial \tau}{\partial t} \tag{115}$$

so that from Equation 113 we have

$$\frac{\partial \xi}{\partial u} = 1, \quad \frac{\partial \xi}{\partial t} = o, \qquad \frac{\partial \tau}{\partial t} = 1 \quad \text{and} \quad \frac{\partial \tau}{\partial u} = -\frac{1}{q} \tag{116}$$

Therefore,

$$\frac{\partial s}{\partial u} = \frac{\partial s}{\partial \xi} - \frac{1}{q}\frac{\partial s}{\partial \tau} \tag{117}$$

and

$$\frac{\partial s}{\partial t} = \frac{\partial s}{\partial \tau} \tag{118}$$

Substituting into Equation 106 gives

$$q\frac{\partial s}{\partial \xi} = -kxs \tag{119}$$

and similarly, from Equation 107,

$$q \frac{\partial p}{\partial \xi} = kxs \tag{120}$$

The problem in terms of substrate and product is therefore reduced to only one independent variable, rather than two. The solution of Equation 119 and 120 is

$$s = S_r \exp(-kxu/q) \tag{121}$$

$$p = kxus/q \tag{122}$$

where S_r is the input nutrient concentration and we have replaced ξ with u. Saunders and Bazin[43] have extended this rather trivial example to a two-stage nitrification process in which ammonium is converted to nitrite by *Nitrosomonas* and nitrite is used by *Nitrobacter* producing nitrate. Two models were analyzed, each based on a different growth rate function. In one model it was supposed that growth was nutrient dependent, while in the other example some other factor was assumed to control the reproductive rate. Using the change of variable technique outlined above they were able to solve each set of equations and so demonstrate the behavior of the system for these two limiting cases.

One further point should be mentioned with respect to microbial activity in packed columns. The models of Prosser and Gray[44] and Saunders and Bazin[43] assume implicitly that the film of biomass adhering to the solid substratum does not become so thick that diffusion of metabolites through it becomes a limiting factor. This assumption is not always tenable, especially when thick layers of slime embedded with microorganisms build up. In such cases diffusion-limited growth terms must be incorporated into models of continuous flow systems.[45-47]

REFERENCES

1. **Monod, J.,** La technique de culture continue; theorie et applications. *Ann. Inst. Pasteur, Paris,* 79, 390, 1950.
2. **Novick, A. and Szilard, L.,** Description of the chemostat, *Science,* 112, 715, 1950.
3. **Herbert, D.,** A theoretical analysis of continuous culture systems, in *Continuous Culture,* Soc. Chem. Ind. Monogr. No. 12, Page Brothers, Norwich, 1964, 21.
4. **Pirt, S. J.,** *Principles of Microbe and Cell Cultivation,* Blackwell, Oxford, 1975.
5. **Powell, E. O.,** Theory of the chemostat, *Lab. Pract.,* 14, 1145, 1965.
6. **Herbert, D., Elsworth, R., and Telling, R. C.,** The continuous culture of bacteria: a theoretical and experimental study, *J. Gen. Microbiol.,* 14, 601, 1956.
7. **Monod, J.,** *Recherches sur la Croissance des Cultures Bacteriennes,* Hermann & Cie, Paris, 1942.
8. **Bazin, M. J., Saunders, P. T., and Prosser, J. I.,** Models of microbial interactions in the soil, *CRC Crit. Rev. Microbiol.,* 4, 463, 1976.
9. **Patten, B. C.,** A primer for ecological modeling and simulation with analog and digital computers, in *Systems Analysis and Simulation in Ecology,* Vol. 1, Patten, B. C., Ed., Academic Press, New York, 1971.
10. **Koga, S. and Humphrey, A. E.,** Study of the dynamic behavior of the chemostat system, *Biotechnol. Bioeng.,* 9, 375, 1967.
11. **Perram, J. W.,** Relaxation times in bacteriological culture and the approach to steady state, *J. Theor. Biol.,* 38, 571, 1973.
12. **Caperon, J.,** Time lag in population growth response of *Isochrysis galvana* to variable nitrate environment, *Ecology,* 50, 188, 1969.

13. **MacDonald, N.**, Time delay in simple chemostat models, *Biotechnol., Bioeng.*, 18, 805, 1976.
14. **Contois, D. E.**, Kinetics of bacterial growth: relationship between population density and specific growth rate of continuous cultures, *J. Gen. Microbiol.*, 21, 40, 1959.
15. **Ramkrishna, D., Fredrickson, A. G., and Tsuchiga, H. M.**, Dynamics of microbial propagation: models considering inhibitors and variable cell composition, *Biotechnol. Bioeng.*, 9, 129, 1967.
16. **Williams, F. M.**, A model of cell growth dynamics, *J. Theor. Biol.*, 15, 190, 1967.
17. **Williams, F. M.**, Dynamics of microbial populations, in *Systems Analysis and Simulation in Ecology*, Vol. 1, Patten, B. C., Ed., Academic Press, New York, 1971.
18. **Williams, F. M.**, Mathematics of microbial populations, with emphasis on open systems, in *Growth by Interssusception*, Deevey, E. S., Ed., Archron Books, Connecticut, 1973, 395.
19. **Dahes, J. N., Finn, R. K., and Wilke, C. R.**, Equations of substrate-limited growth: the case for Blackman kinetics, *Biotechnol. Bioeng.*, 15, 1159, 1973.
20. **Van Dedem, G. and Moo Young, M.**, A model for diauxic growth, *Biotechnol. Bioeng.*, 17, 1301, 1975.
21. **Fredrickson, A. G.**, Formulation of structural growth models, *Biotechnol. Bioeng.*, 18, 1481, 1976.
22. **Knowles, G., Downing, A. L., and Barrett, M. J.**, Determination of kinetic constants for nitrifying bacteria in mixed culture, with the aid of an electronic computer, *J. Gen. Microbiol.*, 38, 263, 1965.
23. **Topiwala, H. and Sinclair, C. G.**, Temperature relationships in continuous culture, *Biotechnol. Bioeng.*, 13, 795, 1971.
24. **Edwards, V. H.**, The influence of high substrate concentrations on microbial kinetics, *Biotechnol. Bioeng.*, 12, 679, 1970.
25. **Yano, T. and Koga, S.**, Dynamic behavior of the chemostat subject to substrate inhibition, *Biotechnol. Bioeng.*, 11, 139, 1969.
26. **Yano, T. and Koga, S.**, Dynamic behavior of the chemostat subject to product inhibition, *J. Gen. Appl. Microbiol.*, 19, 97, 1973.
27. **Ryder, D. N. and Sinclair, C. G.**, Model for the growth of aerobic microorganisms under oxygen limiting conditions, *Biotechnol Bioeng.*, 14, 787, 1972.
28. **Sykes, R. M.**, Identification of the limiting nutrient and specific growth rate, *J. Water Pollut. Control Fed.*, 45, 888, 1973.
29. **Megee, R. D., Drake, J. F., Fredrickson, A. G., and Tsuchiga, H. M.**, Studies in intermicrobial symbiosis, *Saccharomyces cerevesiae* and *Lactobacillus cosei, Can. J. Gen. Microbiol.*, 18, 1733, 1972.
30. **Bader, F. G., Meyer, J. S., Fredrickson, A. G., and Tsuchiga, H. M.**, Comments on microbial growth rate, *Biotechnol. Bioeng.*, 17, 279, 1975.
31. **Bungay, H. R. and Bungay, H. L.**, Microbial interaction in continuous culture, *Adv. Appl. Microbiol.*, 19, 269, 1968.
32. **Veldkamp, H. and Kuenen, J. G.**, The chemostat as a model system for ecological studies, *Bull. Ecol. Res. Comm. (Stockholm)*, 17, 347, 1973.
33. **Veldkamp, H.**, *Continuous Culture in Microbial Physiology and Ecology*, Meadowfield, Co., Durham, 1976.
34. **Curds, C. and Bazin, M. J.**, Protozoan predation in batch and continuous culture in *Advances in Aquatic Microbiology*, Vol. I, Droop, M. R. and Jannasch, H. W., Eds., Academic Press, London, 1977, 115.
35. **Canale, R. P.**, An analysis of models describing predator-prey interaction, *Biotechnol. Bioeng.*, 12, 353, 1970.
36. **Saunders, P. T. and Bazin, M. J.**, On the stability of food chains, *J. Theor. Biol.*, 52, 121, 1975.
37. **Sanchez, D. A.**, *Ordinary Differential Equations and Stability Theory: an Introduction*, W. H. Freeman, San Francisco, 1968, 57.
38. **Topiwala, H. W. and Hamer, G.**, Effect of wall growth in steady-state continuous cultures, *Biotechnol. Bioeng.*, 13, 919, 1971.
39. **Howell, J. A., Chi, C. T., and Pawlowsky, U.**, Effect of wall growth on scale-up problems and dynamic operating characteristics of the biological reactor, *Biotechnol. Bioeng.*, 14, 253, 1972.
40. **Chi, C. T., Howell, J. A., and Pawlosky, U.**, The regions of multiple stable steady states of a biological reactor with wall growth utilizing inhibitory substrates, *Chem. Eng. Sci.*, 29, 207, 1974.
41. **Atkinson, B.**, *Biochemical Reactors*, Pion Press, London 1974, 215.
42. **Bazin, M. J. and Saunders, P. T.**, Dynamics of nitrification in a continuous flow system, *Soil Biol. Biochem.*, 5, 531, 1973.
43. **Saunders, P. T. and Bazin, M. J.**, Nonsteady state studies of nitrification in soil: theoretical considerations, *Soil Biol. Biochem.*, 5, 545, 1973.
44. **Prosser, J. I. and Gray, T. R. G.**, Use of finite difference method to study a model system of nitrification at low substrate concentrations, *J. Gen. Microbiol.*, 102, 119, 1977.

45. **Pirt, S. J.**, A quantitative theory of the action of microbes attached to a packed column relevent to trickling, filter effluent purification, and to microbial action in soil, *J. Appl. Chem. Biotechnol.*, 23, 389, 1973.
46. **Saunders, P. T. and Bazin, M. J.**, Attachment of microorganisms in a packed column: metabolite diffusion through the microbial film as a limiting factor, *J. Appl. Chem. Biotechnol.*, 23, 847, 1973.
47. **Atkinson, B., Busch, A. W., Swilleg, E. L., and Williams, D. A.**, Kinetics, mass transfer, and organism growth in a biological film reactor, *Trans. Inst. Chem. Eng.*, 45, 257, 1967.

Chapter 4

COMPLEX SYSTEMS*

J. Řičica and P. Doberský

TABLE OF CONTENTS

I. Introduction 64

II. Single-Stream Multi-Stage Systems 64
- A. Characteristics of Growth 64
  - 1. Cell Mass Balance 65
  - 2. Substrate Balance 71
- B. Product Formation 75

III. Single-stream Gradient Systems 81
- A. Tubular Fermentor 81
- B. Simulation of Plug-Flow Fermentor 84

IV. Multi-stream Systems 87
- A. Inflow of Further Substrate 87
- B. Inflow of Further Microorganisms 88
- C. Dialysis Fermentation 92
- D. Solid Substrate Fermentation 93

References 94

* This chapter was submitted in March 1978.

I. INTRODUCTION

Continuous cultivation, especially a multi-stage complex system, is an essential experimental tool increasingly attractive for exploring the evolution of a population of microorganisms, not only with the aid of studies in growth physiology and kinetics, metabolism, and a certain product formation, but also in molecular biology, genetics, biochemistry, and ecology in relation to the whole spectrum of microbiological problems.

Owing to the physiological changes, the reaction order of growth usually varies and the kinetics of product formation may be of the same or of a different order from that of the growth kinetics. Multi-stage complex systems might be therefore advantageous and optimal for most microbial processes and giving the lowest total holding time needed to achieve a desired stage of conversion. Although the characteristics of growth and physiological state of the population are basic presumptions for the design of continuous complex systems, their final technological arrangement is determined by the kinetics of a given product formation, by utilization of complex substrates, or by type of metabolism, adaptation, or selection.

The basic theoretical analysis, both of the growth of a population of microorganisms and of a certain product formation in multi-stage complex systems was developed by Herbert,[1,2] Aiba et al.,[3] Fencl,[4] Powell and Lowe,[5] and Bischoff.[6] Further basic information about the design of different reactor types and systems can be found in e.g., Aris,[7] Denbigh and Turner,[8] and Kafarov.[9]

The problem was analyzed in more detail in the above articles than in this chapter, and even particular cases were taken into consideration. In the following text, some ideas, opinions, and equations necessary for characterizaton of basic principles of complex continuous systems are employed following Chapter 2.

II. SINGLE-STREAM MULTI-STAGE SYSTEMS

A. Characteristics of Growth

The performance of continuous cultures can be more or less predicted and derived from batch experimental data by application of a graphical solution based on the increase of the cell mass in time, as it was shown by Luedeking and Piret,[10] Bischoff,[6] Kono,[11] and Kono and Asai.[12,13] The increase of cell mass, i.e., the growth rate, is a complex function of many variables and can be considered as an indicator of the physiological state of the culture of microorganisms. However, graphical estimations and predictions of cell and product concentrations for a continuous cultivation may not be reliable, since the batch data are obtained under continuously varying conditions quite different from permanent steady-state conditions in a continuous culture. Nevertheless, these estimations are not totally worthless; they can be a valuable guide for the design and operation of a continuous culture system. For both the prediction and the analysis of complex culture systems it is necessary to carry out a mathematical treatment of individual cases.

When data obtained in the batch culture are treated graphically, the curve is drawn through individual points that are often rather far from each other, and the values of which are highly scattered. In order to obtain more reliable results, it is possible to utilize a statistical treatment of mean values from cultivations performed under practically identical conditions. A model usually determined by a system of mathematical relations and characterizing primary assumptions about the character of growth and metabolic processes is derived. Considerations concerning these relations will be limited to simple deterministic models, the coefficients of which can be derived from curves of growth, of substrate consumption, and of product formation using regression methods. Regression is a comparison of the course of these curves with that of the model function, the mathematical description of which is known and characterizes the process under study. This comparison is based on certain criteria, the least square method being most commonly used.

It has been known that the equation:

$$\frac{dX}{dt} = \mu X \tag{1}$$

describes the growth of a population of microorganisms. The change "dX/dt" of cell mass concentration, X, at time, t, depends on the specific growth rate, μ. Its maximum value, μ_{max}, is attained when organisms grow exponentially.

The specific growth rate μ, [time^{-1}], is defined as a quantity of cell mass formed by the unit quantity of cell mass per unit time. The value of μ is not constant (see Chapter 2), but depends on the concentration of growth-limiting substrate, S, in the culture, on μ_{max} at saturation levels of the substrate and on the saturation constant, K_s, numerically equal to the substrate concentration, $K_s = S$, at which $\mu = \mu_{max}/2$, as it is defined by Monod's kinetic model:

$$\mu = \mu_{max} \frac{S}{K_S + S} \tag{2}$$

Any substance can be taken as a limiting substrate, but usually a carbon source is considered as growth limiting.

If the dependence of the specific growth rate on the concentration of the limiting nutrient is influenced, e.g., by transport mechanisms, competitive inhibition, true and apparent diffusion, endogenous metabolism, and maintenance relations, Equation 2, is to be modified. This type of influence is discussed in more detail, e.g. by Powell,[14,15] Schulze and Lipe,[16] Contois,[17] Pirt,[18] Van Uden,[19,20] Edwards and Wilke,[21] Aiba et al.,[22] Nikolayev and Sokolov,[23] and Dabes et al.[24] However, in the steady state of a culture grown in the single-stage chemostat, the specific growth rate equals to the dilution rate, $\mu = D$, as shown by Equation 6, and becomes a variable independent on S. Hence, if the culture in the first stage of a series of fermentors is kept under stable conditions, $\mu_{max} > D$, the whole system is stable, too. Since the specific growth rate is a complex function of many variables, it might be considered as an indicator of the physiological state of microorganisms.

Nyholm[25] examined a kinetic model expressing the specific growth rate of algae as a function of the intracellular phosphorus content. Bijkerk and Hall[26] used the cell cycle to suggest the basic structure of a two-stage deterministic model representing the preparatory process of substrate uptake and conversion separately from replication and division. The regulation of the culture fraction developed to each of these broad areas of metabolism and the overall growth rate is related to the nature and availability of the energy substrate. The predicted steady-state values were calculated using a computerized trial-and-error method.

1. Cell Mass Balance

Let us assume that a single-stream multi-stage continuous system consists of a chain of chemostatic fermentors, i.e., of perfectly mixed reactors in series of either equal or unequal volumes, V_1, V_2, ..., V_{n-1}, V_n. The number of stages (fermentors) may be 2 or N. A certain quantity of the culture medium, F, [volume·time^{-1}], is fed to the first stage from which it flows to the next one and so on, so that the actual flow through all stages is identical. The dilution rate, D, [time^{-1}], in individual stages (fermentors) is defined as a ratio of the medium feeding rate and the volume of the culture, $D = F/V$. The reciprocal value of the dilution rate is the holding time, θ, [time], representing the mean time in which a hypothetical particle of the culture resides inside the corresponding stage. As individual dilution rates are fixed ratios, then the actual flow in dependence on culture volumes is:

$$F = D_1 V_1 = D_2 V_2 = \ldots. = D_{n-1} V_{n-1} = D_n V_n \qquad (3)$$

where the subscripts 1,2, n-1, n, indicate the conditions entering and leaving the first, second and n-th stage, respectively. The complexity of a multi-stage system increases with the number of stages.

The concentration of cells, X_n, in an arbitrary n-th stage can be described by the cell mass balance equation: increase = inflow + growth − outflow

$$\frac{dX_n}{dt} = D_n X_{n-1} + \mu_n X_n - D_n X_n \qquad (4)$$

If microorganisms are grown in permanent steady-state conditions, the increase of cell mass with time, dX_n/dt, is zero and the concentration of cells is permanently constant.

Since in the first stage, which behaves like a single-stage fermentor, the inflow of cells, $D_n X_{n-1}$, is zero, the growth equals to the outflow:

$$\mu_1 X_1 = D_1 X_1 \qquad (5)$$

$$\mu_1 = D_1 \qquad (6)$$

and the specific growth rate, μ, is dependent on the dilution rate, D, only. The cell mass balance Equation 4 may be rewritten:

$$\frac{dX_n}{dt} = \mu_n X_n - D_n (X_n - X_{n-1}) \qquad (7)$$

and rearranging Equation 7 for steady-state condition, $dX/dt = 0$, the value of μ_n can be derived:

$$\mu_n = D_n \left(\frac{X_n - X_{n-1}}{X_n} \right) \qquad (8)$$

Thus in the n-th stage the specific growth rate is no more equal to the dilution rate, seeing that it is influenced by the inflow of cells from the preceding stage. Since the portion of inflowing cells, X_{n-1}, is not only the quantitative part of X_n, but is composed also of cells of different physiological state originated in foregoing stages, the sum of cells, X_n, is therefore nonhomogenous. This physiologial nonhomogeneity increases with the number of stages. Nevertheless, the steady-state growth conditions in the n-th stage can be described by the rearranged Equation 7:

$$X_n - X_{n-1} = \mu_n X_n \frac{1}{D_n} \qquad (9)$$

The holding time, $\theta_n = 1/D_n$, needed for cells to reach the concentration, X_n, is:

$$\theta_n = \frac{1}{\mu_n X_n}(X_n - X_{n-1}) \qquad (10)$$

Since in the first stage, $X_{n-1} = 0$, the holding time, θ_1, is given:

$$\theta_1 = \frac{1}{\mu_1 X_1}(X_1 - 0) \qquad (11)$$

The right hand side terms of Equation 11 in the form presented here are suitable for graphical interpretation demonstrated later.

The total holding time, θ_T, of a multi-stage system composed from N fermentors is then:

$$\theta_T = \theta_1 + \theta_2 + \ldots . + \theta_{n-1} + \theta_n \qquad (12)$$

Material balance equations and experimental data obtainable from a batch cultivation usually provide information for a graphical approach to the design of continuous cultivations and to the prediction of their performance characteristics. Although some of growth characteristics expressed by a graphical method are not identical with those obtained in a continuous cultivation, graphical methods described by Luedeking and Piret,[10] Bischoff,[6] Kono,[11] and Kono and Asai,[12,13] can be used as a worthwhile guide for the operation of a multi-stage culture system.

The growth of a population is usually expressed by the equation, $dX/dt = \mu X$. Plotting dX/dt against X, a curve representing the output of the system, DX, is obtained (Figure 1A).

As the entering concentration in the first stage is zero, the slopes of straight lines (1 or 1′) passing through the origin and through selected arbitrary points on the output curve denote dilution rates, $D_1 = F_1/V_1$, in this stage. The selected point represents the output from the first stage, D_1X_1, for the relevant value of X_1 on the abscissa, as can be deduced from Equation 5. In Figure 1A, two examples only are demonstrated with straight lines 1 and 1′. The straight line having the maximum slope and being a tangent to the production curve represents the maximum specific growth rate, μ_{max}.

The entering concentration to the n-th stage is X_{n-1}. The straight line (2) starting at this concentration on the abscissa and intercepting the output curve at a selected point denoting the output, D_nX_n, provides the corresponding concentration of cells, X_n. The slope of the straight line (2) represents the dilution rate, in the n-th stage, D_n, as follows from Equation 9.

In our case a two-stage system is shown. If straight lines (1, 2) are parallel having identical slopes, dilution rates and volumes of both stages are also identical. If different volumes are used, then individual dilution rates are also different; therefore, the slopes of straight lines are not parallel. This case is not demonstrated in Figure 1A.

Using the mathematical analysis of the graphical method, Bischoff[6] employed an optimizing approach to find the shortest holding time, θ, necessary for achieving a desired conversion (Figure 1B). By plotting the reciprocal value $1/(\mu X)$ against X, the reciprocal output curve is obtained.

Since in the single-stage chemostat or in the first stage of a complex system the inflow of cells is zero, then the mean holding time, θ_1, necessary for the population to achieve the concentration, X_1, and the output, D_1X_1, is equal to the area of the rectangle having the base of, $X_1 - 0$, and the height of $1/(\mu_1X_1)$, as shown by Equation 11. If in the first stage the concentration, X_2, is required, the base of, $X_2 - 0$, and the height of $1/\mu_2X_2$, is applied. Employing a two-stage system for the same purpose the growth of cells is distributed into two phases. In the first stage with fast growth the

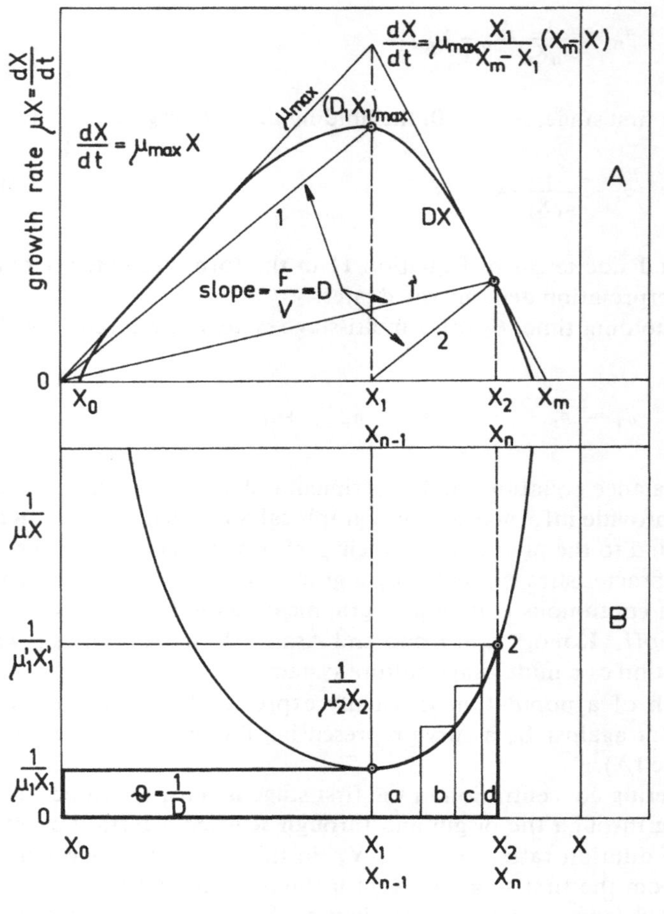

FIGURE 1. Illustration of growth characteristics using methods developed by Luedeking and Piret,[10] and Kono,[11] part A, and Bischoff,[6] part B. X-concentration of cell mass at any point of the production curve DX; X_o — initial concentration of cell mass at zero time; X_1 — concentration of cell mass in the single-stage chemostat or in the first stage of a multi-stage system; DX_{max} — inflexion point between positive and negative acceleration of cell mass increase, where the output of the single-stage chemostat or of the first stage of a multi-stage system is maximum; X_2, X_n — cell mass concentration in the second or in the n-th stage of a multi-stage system; dX/dt, μX — growth rate, [mass·time^{-1}]; μ, μ_{max} — specific growth rate and its maximum value, [time^{-1}]; X_m — theoretical maximum cell concentration determined by the model; F — volumetric flow rate, [volume·time^{-1}]; V — volume of the culture; D — dilution rate, F/V, [time^{-1}]; θ — holding time, a reciprocal value of dilution rate, 1/D, [time]; 1 — slopes of the straight line intersecting the origin and a point of the production curve, DX, represents the dilution rate, D, in the single-stage chemostat or in the first stage of a multi-stage system; the slope of a tangent represents μ_{max}; 2 — the slope of the straight line intercepting the abscissa at concentration X_1 and the production curve at the point 2 represents the dilution rate in the second stage of a multi-stage system, D_2. If the volumes of individual stages are equal the slopes are parallel, if they are unequal the slopes are different.

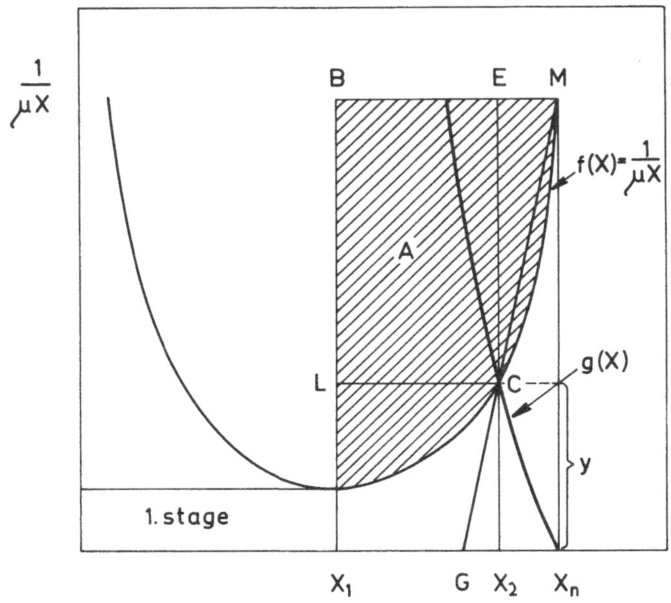

FIGURE 2. The reduction of holding time substituting two smaller stages of equal volumes for one large stage. X_1, X_2, X_n — concentrations of cell mass in corresponding stages; $1/\mu X$ — reciprocal output curve; y — coordinate of the point C; C — point of intersection of curves g(X), and f(X), decisive for the construction of rectangles X_1X_2CL, and X_2X_nME; A — hatched area indicating the difference between holding times of a continuous stirred fermentor and an ideal plug-flow fermentor.

mean holding time, θ_1, is given by the rectangle, $(X_1 - 0)\ 1/(\mu_1 X_1)$, and in the second stage with retarded growth, the mean holding time, θ_2, is represented by the rectangle, $(X_2 - X_1)\ 1/(\mu_2 X_2)$, as substantiate Equations 10 and 11.

Comparing both systems, when used for achieving the same concentration, X_2, it can be seen that the sum of the rectangle areas, i.e., of the mean holding times in the two-stage system, is less than in the single-stage one. In this way the growth process can be distributed into a series of homogenous fermentors with very short holding times, (e.g., areas a, b, c, and d), thus approaching the ideal course of that part of the output curve that would be performed otherwise in an ideal tubular plug-flow fermentor. As Bischoff[6] pointed out, the holding time of an ideal plug-flow fermentor is represented by the area between the curve and the abscissa in the section, X_n-X_{n-1}. The reader's attention is directed to the fact that the ideal area denoting the optimum holding time lies below the output curve, $1/(\mu X)$, and that the parts of rectangles overlapping the curve stand for mean discrepancies in comparison with an ideal situation or with a batch culture. The rectangles shown in Figure 1B, are not an optimum choice; they are casual examples only.

A method presented here might be facilitating the design of a series of fermentors in which the needed concentration of cell mass, X_n, could be achieved. The holding time in the first fermentor is chosen by means of graphical and numerical ways (Equations 11 and 7), in order to correspond very near to the minimum of the reciprocal growth curve, $1/(\mu X) = f(X)$, providing the concentration X_1 (Figure 2). Concentrations higher than X_1 are then obtained, if one or more further fermentors are attached in series. A precondition is given that all those fermentors, except the first one, have identical volumes, i.e., identical holding times represented by relevant areas of rectangles. In order to design these rectangle areas it is necessary to choose the values of corresponding concentrations of $X_2, X_3,..,X_{n-1}$.

In a two-stage system, N = 2, the holding time, θ_2, in the second fermentor is represented by the rectangle, X_1X_nMB, (Figure 2):

$$\theta_2 = \frac{1}{(\mu X)_n}(X_n - X_1) \tag{13}$$

The continuous fermentation of whey lactose to lactic acid developed by Bischoff,[6] and Keller and Gerhardt[27] can serve as an example. The simulation on a digital computer predicted the holding times for experimental test and the use of two stages for the greatest efficiency.

If a three-stage system is intended to be designed, N = 3, the concentration difference, $X_n - X_1$, is to be divided into two unequal parts providing bases of rectangles, X_1X_2CL, and X_2X_nME, in order to obtain equal areas. (Figure 2). From the equality of rectangles it follows

$$(X_2 - X_1)\frac{1}{(\mu X)_2} = (X_n - X_2)\frac{1}{(\mu X)_n} \tag{14}$$

describing the curve g(X). Now the problem is arising how to establish the point, C, determining the concentration, X_2, and being decisive for the construction of both rectangles. Since the point of intersection, C, of the curve, g(X), with the curve, $1/(\mu X) = f(X)$, is defined by coordinates, X_2, and y, the curve g(X), can be specified by parametric equations:

$$y = \frac{1}{(\mu X)_2} = k\frac{1}{(\mu X)_n} \tag{15}$$

$$X_2 = \frac{kX_1 + X_n}{k + 1} \tag{16}$$

where the parameter, k, lies within the range, $0 \leqslant k \leqslant 1$.

Each of the rectangles, X_1X_2CL, and X_2X_nME, can be again divided into two further smaller ones. Every division results in a significant reduction of the holding time, i.e., of the rectangle areas in comparison with the initial one before the division. The example in Figure 2 shows the decrease of the holding time represented by the area, A_1, of the rectangle, BLCE:

$$A_1 = (X_2 - X_1)\left[\frac{1}{(\mu X)_n} - \frac{1}{(\mu X)_2}\right] \tag{17}$$

Substituting, X_2, and $1/(\mu X)_2$, from Equations 15 and 16 for Equation 17, we have:

$$A_1 = (X_n - X_1)\frac{1}{(\mu X)_n}\frac{1-k}{1+k} \tag{18}$$

The area, A, (Figure 2 hatched) above the curve, f(X), indicating the difference between holding times of a continuous stirred fermentor and a tubular plug-flow one, is always smaller than the area, A_2, of the tetragon, X_1GMB, $A < A_2$:

$$A < A_2 = \frac{1}{2(\mu X)_n}\left(X_n + \frac{kX_n - X_2}{k-1} - 2X_1\right) \quad (19)$$

$$A < A_2 = (X_n - X_1)\frac{1}{(\mu X)_n}\frac{k^2 + \frac{k}{2} - 1}{(1+k)(1-k)} \quad (20)$$

The decrease of the area difference, A, when divided into two smaller stages is characterized by the ratio, A_1/A, which is not smaller in any case than the ratio A_1/A_2:

$$\frac{A_1}{A} > \frac{A_1}{A_2} = \frac{(k-1)^2}{1 - \frac{k}{2} - k^2} \quad (21)$$

If the extreme of this expression is estimated, the ratio A_1/A_2 attains the minimum value at $k = 3/5$, for which

$$\frac{A_1}{A} > \frac{A_1}{A_2} = \frac{8}{17} \quad (22)$$

Equation 22 signifies that the division of one stage into two smaller ones reduces the holding time at least by about (8/17) 100% of the initial value, i.e., almost about one half.

The dependence of the holding time on the number of stages in series succeeding the first stage, designed by the method of twofold increase of the number of smaller stages, N = 2, 4, 8...., is shown in Figure 8, curve 2. The curve 2 also indicates the upper limit that cannot be overpassd by differences of holding times distinguishing from an ideal case, θ_i, even with the most unfavorable course of the curve $f(X) = 1/(\mu X)$.

In a case where the object is to produce the cell mass only, using a simple medium, there is little to be gained by the application of two or more stages in comparison with a single fermentor, which is equally efficient. Complex systems may reveal their advantages when the substrate, even a multiple one in a complex medium, is to be effectively consumed or when it is necessary to cultivate cells in a particular physiological state, i.e., cells having a particular quality and ability to produce a product of interest.

2. Substrate Balance

Microorganisms increase the cell mass with time, $dX/dt = \mu X$, utilizing the substrate, S, [mass·volume^{-1}], simultaneously:

$$-\frac{dS}{dt} = qX \quad (23)$$

where, q, [time^{-1}], is the specific rate of consumption, $q = q_{max}S/(K_s + S)X$, which was discussed by Powell[14] as a "metabolic coefficient". Since both the metabolic coefficient and the composition of cells are functions of specific growth rate, the yield of cell mass, is variable with μ or D. The yield of cell mass, i.e., the yield coefficient, Y, is the mass of cells produced per unit mass of a particular substrate consumed:

$$-\frac{dX}{dS} = \frac{\mu X}{qX} = Y \quad (24)$$

The metabolic coefficient is then given by the ratio:

$$q = \frac{\mu}{Y} \quad (25)$$

Although in Equation 24 the yield coefficient, Y, is defined with respect to variable concentration of the limiting substrate, it is not a function of this particular substrate only. In order to make a mathematical description of the yield coefficient more accurate, it could be necessary to discriminate among different types of metabolism and consider the effect of concentration of the medium components including metabolites inhibiting the growth. Herbert[28] mentioned an equation in which the effect of endogenous metabolism, μ_e, and of the dilution rate, D, on the observed yield coefficient, Y, may be expressed as, $Y = Y_g D/(D - \mu_e)$, where Y_g is the yield that would be obtained in the absence of endogenous metabolism.

Herbert[2] has also shown that at comparable dilution rates the utilization of the substrate is more efficient in a multi-stage system than in the single-stage system, i.e., the effective yield is higher.

The balance equations for the substrate, in both the n-th stage and the first one are identical; the subscripts are to be changed only:

increase = input − output − consumption

$$\frac{dS_n}{dt} = D_n S_{n-1} - D_n S_n - q_n X_n \quad (26)$$

In steady conditions $dS_n/dt = 0$ and we can write:

$$(S_{n-1} - S_n) = q_n X_n \cdot \frac{1}{D_n} \quad (27)$$

The holding time, θ_n, i.e., the reciprocal value of dilution rate, $1/D_n$, is

$$\theta_n = (S_{n-1} - S_n) \frac{1}{q_n X_n} \quad (28)$$

Equation 27 can be rewritten by substituting, q, from Equation 25:

$$(S_{n-1} - S_n) = \frac{\mu_n}{Y_n} \frac{1}{D_n} X_n \quad (29)$$

Calculating, $X_n = (X_n - X_{n-1}) D_n/\mu_n$, from Equation 9, and rearranging Equation 29, one may deduce the simple expression:

$$Y_n (S_{n-1} - S_n) = X_n - X_{n-1} \quad (30)$$

sustaining the steady condition that the concentration of cell mass is proportional to the difference between entering and leaving substrate concentrations owing to the yield coefficient.

As the rate of the substrate consumption is described by, $-dS/dt = qX$, then by plotting the reciprocal value, $1/(qX)$, against S, a reciprocal consumption curve similar to the reciprocal cell output curve may be obtained. Equation 28 shows that the holding time, $\theta_n = 1/D_n$, in the n-th stage, needed for the consumption of the substrate, S_{n-1}, down to S_n, and for the increase of cell mass up to X_n, is represented by the rectangle, $(S_{n-1} - S_n)\ 1/(q_nX_n)$. The shape of the consumption curve is similar to the cell output curve, but the course in converse. Since the points of inflexion of both curves lie at the same X, and the growth of cells is stopped at a time when the substrate is exhausted, the graphical construction of holding times from reciprocal consumption curve is usually not necessary. This interpretation should be used in such cases only, when microorganisms are grown in dense media containing solid particles, and the estimation of X is rather difficult.

In media containing multiple carbon sources the effect on enzyme synthesis and activity may occur. These catabolite repression and catabolite inhibition give a growth pattern called the polyauxic effect. Since in the continuous culture the steady-state concentrations of substrates are low, the growth pattern is distinct from that of batch culture, and the effect of both catabolite repression and inhibition may be significantly reduced. Van Dedem and Moo-Young,[29] Tsao and Yange,[30] and Yoon et al.[31] discussed this competitive effect of mixed substrates in the single-stage system in more detail. Sinclair and Ryder[32] developed a steady-state interacting double substrate kinetic model with an endogenous metabolism term explaining continuous culture data obtained for the growth of yeast. Cooney and Wang[33] examined dual nutrient limitation of nitrogen and phosphate and the transient response of the culture following a pulse of ammonia. The nitrogen limitation restricted protein synthesis, while the phosphate limitation restricted nucleic acid synthesis. Single-stage operation, while possible, is not necessarily the best solution. The effective utilization of multiple substrate will be easier in a multi-stage system.

Dilution rates used in the first stage (Figure 1), are slower than the maximum specific growth rate, $D<\mu_m$. The lower the value of D the larger the area of rectangles, $1/(\mu X)$ $(X-0)$. It means that the population is not allowed to pass throughout certain developmental periods. Individual cell cycles take place in permanent steady conditions in contrast to variable conditions in the batch. Hence, the physiological characteristics differ from that in the batch and this difference may be considered as a rectangle area overlapping the output curve.

The long term cultivation of some microorganisms, especially filamentous and sporulating, under steady-state conditions can result in an undesirable performance such as "degeneration", "mutation", and "failure of production capacity". Variations in metabolic stability owing to different inhibitory and competitive effects may manifest in oscillations.

The minimum rectangle areas, which do not overlap the output curve, represent holding times, i.e., dilution rates, around the point of inflection, where the output of cell mass is maximum, $(DX)_{max}$, as shown in Figure 1. Only the physiological state of the population grown continuously in these conditions is agreeable with that one of the corresponding growth period of the batch culture. It has been known that both the yield coefficient and the composition of cells are changed when microorganisms are grown on carbon or nitrogen sources as limiting nutrients. They are also variable with dilution rate, as presented by Herbert.[28] Nevertheless, with dilution rates near to $(DX)_{max}$, both the composition and the yield of cells are stable and keep the same value even when cells are grown on different limiting substrates. The growth rate and the metabolic activity become nonlimited by the concentration and chemical character of nutrients. Specific rates of DNA, RNA, and cell protein synthesis may attain maximum values, as mentioned by Řičica.[34] Therefore, the conditions suitable for $(DX)_{max}$ in the first stage should be used, if the development of a population or the formation of a product in a complex system is explored. However, for the sake of safety operation it is better to choose a somewhat lower dilution rate than that providing $(DX)_{max}$. The highest specific growth rate, μ_1, being reduced by the term $S_1/(K_s + S_1) < 1$ (Equation 2), is always less than the maximum specific growth rate theoretically possible.

The operation of the first continuous stirred fermentor is based on the course of experimentally determined curve, $1/(\mu X) = f(X)$, the minimum of which represents the optimum dilution rate providing $(DX)_{max}$. The steady-state growth of microorganisms is described by Equation 5. The substrate utilization resulting in cell mass and product formation may be described by the following relation:

$$\frac{dS_1}{dt} = \frac{\mu_1 X_1}{Y_{X/S}} - \frac{\mu_1 X_1}{Y_{P/S}} + D_1 (S - S_1) \tag{31}$$

where S, is the substrate concentration entering the first stage, $\mu_1 = \mu_{max} S_1/(K_s + S_1)$ and $Y_{X/S}$, $Y_{P/S}$ are yield coefficients for cells and product. Using Equations 5 and 31, steady-state concentrations of cells, X_1, and of the substrate, S_1, for $D_1 < \mu_{max}$ can be derived:

$$X_1 = Y \left(S + \frac{D_1 K_S}{D_1 - \mu_{max}} \right) \tag{32}$$

$$S_1 = \frac{D_1 K_S}{\mu_{max} - D_1} \tag{33}$$

where $Y = Y_{X/S} Y_{P/S} (Y_{X/S} + Y_{P/S})$.

The solution of Equations 32 and 33 can be employed for the calculation of the maximum value of the function, $D_1\overline{X}_1 = f(D_1)$, assuming, $D_1 = D_{opt}$. For this optimum dilution rate the function, $D_1\overline{X}_1$, attains the extreme value, when:

$$\frac{d(D_1 \overline{X}_1)}{d D_1} = 0$$

Solving this with respect to Equations 32 and 33 we have:

$$D_{opt} = \mu_{max} \left(1 - \sqrt{\frac{K_S}{K_S + S}} \right) \tag{34}$$

from which it can be made sure that the course of the function, $D_1\overline{X}_1$, has its maximum value D_{opt}. The steady-state concentrations, $\overline{X}_1$, and $\overline{S}_1$, at this, D_{opt}, are predicted:

$$\overline{X}_1 = Y \left[S + K_S - \sqrt{K_S (S + K_S)} \right] \tag{35}$$

$$\overline{S}_1 = \sqrt{K_S (S + K_S)} - K_S \tag{36}$$

The calculations of, D_{opt}, $\overline{X}_1$, and $\overline{S}_1$, are indeed dependent on characteristics of both the culture and of the substrate, i.e., on constants, μ_{max}, K_S, and coefficients, $Y_{P/S}$ and $Y_{X}/_{S}$, which are to be estimated from experimental batch data, obtained around the point of inflection.

Johnson and Berthouex[35,36] and Williamson and McCarty[37] presented experimental methods available for obtaining data to estimate the biokinetic parameters much more precise than those obtained using conventional methods.

B. Product Formation

The increase of the product concentration, P, [mass volume^{-1}], depends on the activity and on the concentration of microorganisms, X, [mass volume^{-1}]:

$$\frac{dP}{dt} = pX \tag{37}$$

The activity of cells may be represented by the specific production rate, p, [time^{-1}], defined as a quantity of the product produced by unit quantity of cells per unit time:

$$p = \frac{dP}{dt}\frac{1}{X} \tag{38}$$

The specific production rate being a complex function of many interrelated factors, among which cells and substrate are essential, p = p (X, S, ...), can be expressed also in other possible adequate ways corresponding to particular processes.

Microorganisms consume the limiting substrate and produce the cell mass with a yield:

$$Y_{X/S} = \frac{\mu}{q} \tag{39}$$

and a product with a yield:

$$Y_{P/X} = \frac{p}{\mu} \tag{40}$$

The yield of the product based on the same substrate is

$$Y_{P/S} = \frac{p}{q} \tag{41}$$

Following relations can be derived from the above equations:

$$Y_{P/S} = Y_{X/S}\,Y_{P/X} \tag{42}$$

$$Y_{X/S} = \frac{Y_{P/S}}{Y_{P/X}} \tag{43}$$

The yield of product, $Y_{P/S}$, is usually expressed in mass units or molar units. Herbert[28] preferred the expression in gram-atom of carbon or nitrogen of the product per gram-atom of carbon or nitrogen of the substrate used, because that application allowed the stoichiometric calculations. The identical expression may be applied for the cell mass yield, $Y_{X/S}$, too.

The consumption rate of the substrate by microorganisms producing a product is represented by a "metabolic coefficient", q, which may be newly defined using Equations 39 and 43:

$$q = \mu \frac{Y_{P/X}}{Y_{P/S}} \tag{44}$$

Substituting, q_n, in Equation 26 by the right hand side term of Equation 44, the mass balance equation for the substrate in the n-th fermentor in series may be rewritten:

$$\frac{DS_n}{dt} = D_n (S_{n-1} - S_n) - \mu_n X_n \frac{Y_{(P/X)_n}}{Y_{(P/S)_n}} \tag{45}$$

As follows from the above definitions, the yield of product is dependent on μ and consequently it is variable with D. On the other hand, the specific growth rate, μ, may be retarded by the increasing concentration of some products.

A linear decrease in μ caused by a toxic metabolite was shown by Dean and Hinshelwood,[38] and Dean et al.[39] An exponential relationship was found by Aiba et al.[40] Ierusalimski[41] reported that a rectangular hyperbola provided the best fit. It is evident that the type of growth rate retardation by a product concentration depends on the chemical structure and on the site and mechanism of the product interference.

As with cells and substrate, the balance equation for the product, P, [mass volume^{-1}], in the n-th fermentor in series may be expressed:

increase = inflow + production − outflow

$$\frac{dP_n}{dt} = D_n P_{n-1} + p_n X_n - D_n P_n \tag{46}$$

The production rate in the n-th fermentor in steady-state, where $dP_n/dt = 0$, is

$$p_n X_n = D_n (P_n - P_{n-1}) \tag{47}$$

The increase of the product concentration from P_{n-1} up to P_n is given by the following equation:

$$P_n - P_{n-1} = p_n X_n \frac{1}{D_n} \tag{48}$$

with corresponding holding time, $\theta_n = 1/D_n$,

$$\theta_n = (P_n - P_{n-1}) \frac{1}{p_n X_n} \tag{49}$$

Since in the first stage the inflowing product concentration is zero, $P_{n-1} = 0$, the holding time, θ_1, needed for the concentration P_1 may be defined:

$$\theta_1 = (P_n - 0) \frac{1}{p_n X_n} \tag{50}$$

From Equation 47 it can also be derived that in the first stage the production equals to the output:

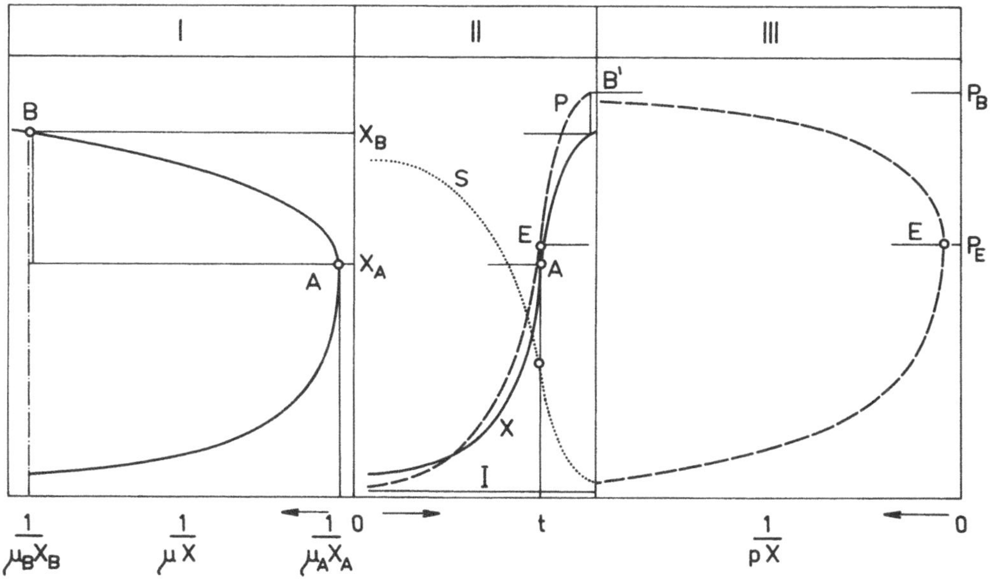

FIGURE 3. Simultaneous accumulation of product and cell mass. X, P, I, S — cell mass, product, intermediate and substrate concentrations; $1/\mu X$ — reciprocal growth rate; 1/pX — reciprocal production rate; A, E — points of inflection at which the culture possesses the maximum activity; B, B′ — points at which 96% of maximum concentrations of both cells and product is achieved; t — time.

$$p_1 X_1 = D_1 P_1 \tag{51}$$

The rate of the product formation obtained from the batch culture, dP/dt = pX, can also be used for the graphical illustration. By plotting the reciprocal value, 1/pX, against P, the reciprocal production curve may be constructed (Figure 3 and Figure 4). Individual holding times required for achieving the chosen product concentration are represented by rectangles defined for the first fermentor by Equation 50, and for the n-th fermentor in series by Equation 49.

Since the product formation depends on the concentration of active cells, holding times providing optimum development of the culture resulting in maximum product yield are to be preferentially derived from the reciprocal growth output curve. Subsequent holding times in which the maximum product accumulation is looked for may then be derived from the corresponding period of the reciprocal product output curve.

According to the kinetics of product formation in a batch culture, it may be considered to what extent the formation of a product is associated with particular developmental periods of the population as a result of energetic or secondary metabolism. Let us assume that there are two main types of product formation.

1. The concentration of the product increases simultaneously, directly with cell mass, $Y_{P/X} = pX/(\mu X)$, and maximum concentrations of both the organisms and product are achieved at the same time (Figure 3). Also inflection points A and E occur at the same time, $t_A = t_E$, (Section II). It is evident that holding times can be derived from the reciprocal growth curve, $1/(\mu X)$, (Section I) Equations 11 and 10, and that the high cell and product concentrations, B, and B′, are achieved already even in a single-stage system, as it is represented by the rectangle $\theta_B(X_B - 0)\ 1/(\mu_B X_B)$. This type of fermentation can be operated continuously with little difficulty and greatly improved efficiency. From the point of view of physiological culture development and rational operation the single-stage cultivation, while

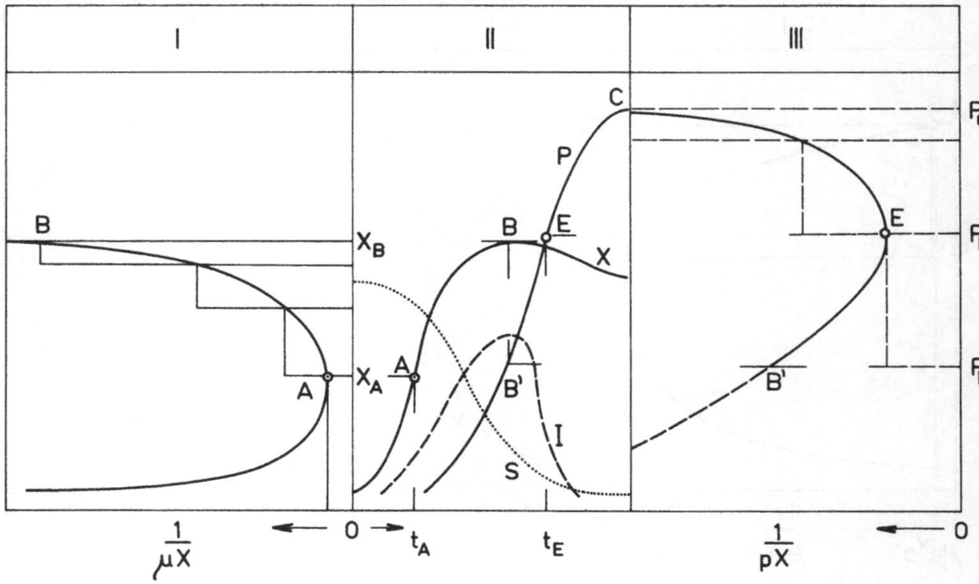

FIGURE 4. Maximum product concentration achieved after the retarded cell mass increase. $1/\mu X$ — reciprocal growth rate; $1/pX$ — reciprocal production rate; t — time; X, P, I, S, — cell mass, product, intermediate and substrate concentrations; A — inflection point of growth rate at which the primary metabolism is maximum and the product concentration is still low; B — 96% of maximum cell mass concentration, X_B; B′ — product concentration $P_B{}'$ achieved during growth period between A and B when projected from B; C — 96% of maximum product concentration, P_C; E — inflection point of production rate with maximum production activity.

possible, is not necessarily the best technique. It may often be more efficient to distribute the process into more than one stage, e.g., into two stages. The dilution rate in the first stage should then be chosen near $(DX)_{max}$ with a corresponding holding time, $\theta_1 = (X_A - 0)\ 1/(\mu_A X_A)$, and in the second stage with holding time $\theta_2 = (X_B - X_A)\ 1/(\mu_B X_B)$. The total holding time required for achieving the concentration B in a two-stage system will then be much lower than the holding time, θ_B, of the single-stage one, $\theta_T < \theta_B$. Production of some organic acids and solvents, vitamins, enzymes, sorbose, lipids, polysaccharides, steroids, transformations of chemicals, etc., can exemplify this type of product information.

2. The product is accumulated throughout the period of retarded cell mass increase and the maximum product concentration is achieved after the growth had stopped (Figure 4), for example the production of some antibiotics, enzymes, vitamins, spores, etc. and multiple substrate utilization.

It is evident that from the point of view of the operational technique and culture development it will be more efficient to separate the growth and product formation. The period of cell mass increase from A to B (Section I) should be distributed into two or three stages. Holding times may be derived from the reciprocal cell output curve (Equations 11 and 10). Since such processes involving the biosynthesis of complex molecules from simpler ones are usually endergonic and, hence, necessarily growth linked, the product is partly formed already during the growth period, but only up to the concentration B′ or P_B (Sections II and III). The inflection point of the production rate, E, however, is distant from the inflection point of the growth rate, A, $t_A \neq t_E$ (Section II), and in some cases even out of the growth period. The production of a secondary product cannot therefore be performed in the single-stage system. Consequently, optimum holding times for $(DP)_{max}$ and product accumulation are to be de-

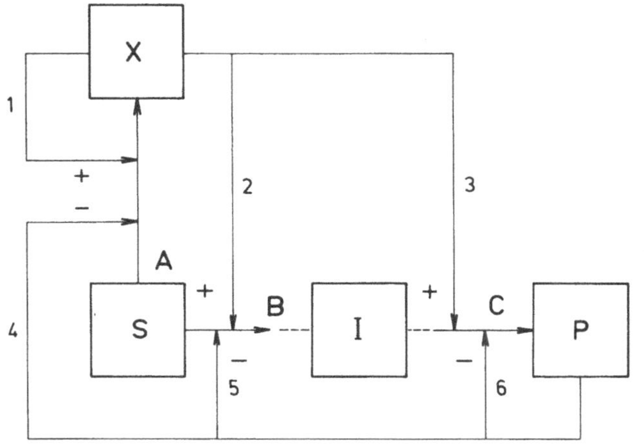

FIGURE 5. Schematic model of product formation. X, P, I, S, — cell mass, product, intermediate, and substrate concentrations; A, B, C, — pathways of mass conversion; 1 to 6, — controlling mechanisms (transfer of information) having positive (+) and negative (−) effects at the site of action.

duced by repeating the procedure starting from B′ and proceeding along the reciprocal production curve (Section III).

When the culture is to be distributed into several stages, it is desirable that the key reactions should take place in a particular stage with the maximum rate, so that the individual holding times should correlate with this demand, as reported by Řičica.[34] In individual stages (fermentors), particular metabolic reactions can be stimulated or restricted by controlling the temperature, pH, partial pressure of O_2 and CO_2, precursor concentration, etc. Some modifications of the medium composition may also be useful. Failures in the continuous production of industrially significant products are often due to insufficient knowledge of biosynthetic pathways controlling the product formation. These are difficult processes to be operated by means of the continuous cultivation, although some antibiotics have been produced by continuous multistage processes on the laboratory scale. It appears likely that there is no general approach that can be applied, and that each product of interest requires a special study before an efficient continuous process can be designed.

A simple schematic model is presented (Figure 5) in which interrelations between cell growth, substrate utilization, product, and intermediate formation are shown. Using this model both types of product formation related to cell mass accumulation, the simultaneous and the delayed one can be described. The pathways A, B, and C represent the conversion of the substrate, S, to the cell mass, X, to the intermediate(s), I, and to the product, P. Controlling mechanisms 1 to 6, acting at different sites, are divided into two groups: (a), positive +, stimulating or supporting the synthetic pathway, and (b), negative −, retarding, restricting, or blocking a given pathway.

In a microbial culture approximated by the model, the controlling mechanisms, 1 to 6, manifest with different intensity. Issuing from the knowledge of the process some regulatory mechanisms can be eliminated, and models less complex may be obtained. In the model presented here, for example, feedback effect of the intermediate, which is assumed not to be significant, has been purposely eliminated. On the other hand, the positive regulatory mechanisms, 1, 2, and 3, initiated by cells, X, are considered to be fully performed. The negative regulatory mechanisms, 4, 5, and 6, evoked by the product, P, denote special interactions. For example, the inhibition of cell growth by the product concentration, 4, results in an indirect restriction of the product synthesis. The direct inhibitory effects, which may operate at different sites of the product synthetic pathway, are demonstrated by the negative interactions, 5 and 6.

In the following set of differential equations the main pathways A, B, and C, are represented by f_A, f_B, and f_C respectively. The controlling mechanisms are described by relevant coefficients, K, considered as variables.

$$\frac{dX}{dt} = f_A(S)\cdot K_1(X)\cdot K_4(P) \tag{52}$$

$$\frac{dI}{dt} = f_B(S)\cdot K_2(X)\cdot K_5(P) - f_C(I)\cdot K_3(X)\cdot K_6(P) \tag{53}$$

$$\frac{dP}{dt} = f_C(I)\cdot K_3(X)\cdot K_6(P) \tag{54}$$

$$\frac{dS}{dt} = -f_A(S)\cdot K_1(X)\cdot K_4(P) - f_B(S)\cdot K_2(X)\cdot K_5(P) \tag{55}$$

In the recent literature, a number of examples of other possible analytical expressions of functions, f, describing the consumption and, K, describing regulation, can be found.

Considering the quantitative differences in positive constants various types of product formation may be approximated by the model. If $f_B < f_C$, the intermediate is utilized immediately and its accumulation does not occur, $I \rightarrow 0$. A similar situation arises when the intermediate is not synthesized at all and the product is formed directly from the substrate, $f_B = 0$, $I = 0$. The delay of product synthesis is negligible, and the accumulation of the product can proceed simultaneously with cell mass accumulation, (Figure 3).

If $f_B > f_C$, the intermediate is temporarily accumulated until the decay of the substrate is completed (Figure 4). The growth rate simultaneously decreases towards zero, $f_A \rightarrow 0$, and the relation between the constants becomes inverse, $f_B < f_C$, due to the utilization of the intermediate for the product formation. Hence, the synthesis of the product is delayed with respect to the growth, and the product accumulation is stopped at about the same time at which the intermediate is almost exhausted.

If in a simplified form of the model the negative regulatory mechanisms, 4, 5, and 6 are omitted, and the positive ones, 1, 2, and 3 initiated by microorganisms are considered in the sense of Monod's kinetics, a set of differential equations describing one of the possible ways of product synthesis may be developed:

$$\frac{dX}{dt} = \mu X = \mu_{max} \frac{SX}{K_S + S} \tag{56}$$

$$\begin{aligned} \frac{dI}{dt} &= iX\, Y_{I/S} - \frac{1}{Y_{P/I}}\, pX \\ &= i_{max} \frac{SX}{K_I + S}\, Y_{I/S} - \frac{1}{Y_{P/I}}\, p_{max} \frac{IX}{K_P + I} \end{aligned} \tag{57}$$

$$\frac{dP}{dt} = pX\, Y_{P/I} = p_{max} \frac{IX}{K_P + I}\, Y_{P/I} \tag{58}$$

$$\frac{dS}{dt} = -\frac{1}{Y_{X/S}}\,\mu X - \frac{1}{Y_{I/S}}\,iX$$
$$= -\frac{1}{Y_{X/S}}\,\mu_{max}\,\frac{SX}{K_S + S} - \frac{1}{Y_{I/S}}\,i_{max}\,\frac{SX}{K_I + S} \qquad (59)$$

The terms, $Y_{X/S}$, $Y_{I/S}$, and $Y_{P/I}$, are corresponding yield coefficients. Saturation constants, K_S and K_I, denote substrate concentrations at which the specific rates of growth and of intermediate synthesis are $\mu = (1/2)\mu_{max}$ and $i = (1/2)\, i_{max}$. The saturation constant, K_P, denotes the concentration of the intermediate at which the specific rate of product synthesis is $p = (1/2)p_{max}$.

For the sake of easier interpretation, the above mentioned differential equations (Equations 52 to 59) are demonstrated in a form describing the batch process. Balance equations describing a continuous cultivation are to be completed by relevant input and output terms.

III. SINGLE-STREAM GRADIENT SYSTEMS

Owing to the physiological changes during the growth of a culture of microorganisms, the growth reaction order usually varies from the autocatalytic first to ordinary first order. If the growth rate or production rate increase with concentration, the optimal output per unit volume of the continuous stirred fermentor is higher than that of the plug-flow type. If the growth rate or production rate decrease with concentration, the continuous plug-flow fermentor is superior to a continuous stirred one, operated for the same period. With zero order reactions, the effectivity of both types is equal. Therefore, for most microbial processes the combination of a continuous stirred fermentor with a plug-flow one should give the shortest total holding time required to achieve the desired concentration of cell mass and product or the required conversion of the substrate. Already Herbert[1] and Powell and Lowe[5] showed the effectivity of a plug-flow fermentor permanently seeded.

A. Tubular Fermentor

An ideal plug-flow fermentor is a continuously operating reactor in which there is a steady movement of infinitesimal portions of the microbial population in a chosen spatial direction. The portion representing a hypothetical culture moves as a piston, and no attempt is made to induce axial mixing between elements of the culture at different points along the direction of flow. The time, θ, for a given element of the culture to flow a distance, L, [m], along the tube or the time necessary for a desired conversion is:

$$\theta = \frac{AL}{F} = \frac{V}{F} = \frac{1}{D} \qquad (60)$$

where A is the cross-sectional area, [m^2].

The growth and metabolic activity of the culture element moving inside the tube result in spatial changes of concentrations of all components in both positive (increase of cell, intermediate, and product concentrations, etc.) and negative (decrease of intermediate and substrate concentrations, lysis, and death of cells, etc.) directions. Changes not only in concentrations, but also in the culture evolution occur. It can be said that continuous gradients not only in concentrations of all culture components, but mainly in the physiological state originate along the direction of flow.

The plug-flow fermentor is usually used for anaerobic processes such as ethanol or lactic acid fermentation. Continuous manufacture of yogurt developed by Driessen et al.[42,43] may serve as an example of a combined two-stage system. Prefermented milk in the first stage is mixed with acidifying milk by means of a centrifugal distributor and brought into a plug-flow fermentor where the coagulation takes place. But even in an anaerobic cultivation in the medium of Newtonian character, the sedimentation of cells and axial and radial irregular mixing caused by bubbles and by flow instabilities cannot be prevented. Hence, an ideal plug-flow fermentor becomes a non-ideal plug-flow imperfectly mixed one. The fraction of the tube volume occupied by gas, R_G, depends upon bubble velocity and gas generation rate and can be obtained as a ratio between superficial gas velocity, U_G,[m· time^{-1}], and bubble velocity, U_B,[m· time^{-1}], as described by Wick and Popper[44] for the grape juice fermentation:

$$R_G = \frac{V_G}{V} = \frac{F_G}{U_B A} = \frac{U_G}{U_B} \qquad (61)$$

where V_G is the gas volume,[m^3], and F_G is the gas volumetric flow rate, [m^3 · time^{-1}]. The residence time of a culture element inside the tube is then reduced

$$\left(\frac{V}{F}\right)_{red.} = (1 - R_G)\frac{V}{F} \qquad (62)$$

In order to achieve the same required conversion, it is necessary (due to the influence of irregularities of flow and of gas hold-up) to modify either the medium flow or prolong the tube.

The main disadvantage of a plug-flow fermentor when used for aerobic cultivation, is the difficulty with which an adequate gas supply throughout the system is achieved. In order to ensure a sufficient oxygen transfer and reduce axial mixing, the internal space of the tube is divided into compartments thoroughly agitated by impellers. The plug-flow fermentor becomes a multi-stage one, and the continuous gradient is transformed to a gradient of stepwise character. Many examples of these fermentor types can be found in the literature, e.g., Greeshields and Smith[45] reviewed the use of these in industrial fermentations. Štěrbáček[46] presented general rules for the optimization of tubular fermentation systems and summarized design methods for the gas hold-up, pressure drop, drop-size distribution, actual interfacial area in nonideal flow conditions, and the prediction of the coefficient of oxygen transfer and the determination of kinetic relations. Completely mixed microbial film fermentors, described for instance by Atkinson and Knights,[47] Howell and Atkinson,[48] and Moser[49] are special cases of heterocontinuous systems coming forward in recent years.

It can be said that the so called plug-flow, piston-flow, tubular, multicompartment tower, and multi-stage fermentors operating in vertical, slant, or horizontal positions can be classified as "gradient fermentors" or "gradient systems".

Let us consider a combined system of an ideal plug-flow fermentor permanently seeded by the concentration, X_1, leaving the continuous stirred fermentor (Figure 1). In the optimum case, the latter operates near the dilution rate providing the maximum output, $(D_1X_1)_{max}$. The plug-flow fermentor should function in such a manner, that throughout the holding time, θ, the development of the culture between concentration X_1 and X_n may occur, and the required concentration X_n would leave the system. The holding time is then given by the time interval necessary for a particle to pass from one end of the tube to the other one, $\theta = t_n - t_1$.

The volumetric flow rate, F, inside the tube is

$$F = A \cdot v \qquad (63)$$

$$v = \frac{F}{A} \tag{64}$$

where v is the velocity of the movement of a particle along the direction of flow, $[m \cdot s^{-1}]$. Throughout the time interval, $t_n - t_1 = \theta$, the particle moving with a velocity, v, travels a distance, which determines the necessary length of the tube, L:

$$L = \theta \cdot v \tag{65}$$

and substituting from Equation 64:

$$L = \theta \frac{F}{A} \tag{66}$$

When cultivating in a tube, the plug-flow is usually destroyed and the ideal length, L_i, must be extended to the real length, L_r. As Kafarov[9] pointed out, the effectiveness, η, of the real tubular fermentor is given by the ratio of lengths or holding times of ideal and real tubular fermentors:

$$\eta = \frac{\theta_i}{\theta_r} \cdot 100\% = \frac{L_i}{L_r} \cdot 100\% \tag{67}$$

In other words, if in an ideal plug-flow fermentor the part of the growth period from X_1 to X_n might have been performed throughout the length L_i, then inside a nonideal (real) fermentor this part should occur along the length L_r; $L_r > L_i$, and $L_r = L_i/\eta$

Though the lengths are in the ratio of L_i/L_R, the specific growth rates μ_i and μ_r for an arbitrary, X, will be in the inverse relation:

$$\frac{\mu_r}{\mu_i} = \frac{L_i}{L_r} = \eta \tag{68}$$

the coefficient η being lower than one, $\eta < 1$. It is shown in Figure 1 and Figure 6 that the ideal holding time, θ_i, required for the increase of concentration from X_i up to X_n is represented by the area below the reciprocal output curve, $f_i(X) = 1/(\mu_i X)$, and between X_1 and X_n:

$$\theta_i = \int_{X_1}^{X_n} \frac{1}{\mu_i X} \, dx \tag{69}$$

The area for a nonideal tubular fermentor lies again in the range between X_1 and X_n, but below the curve, $f_r(X) = 1/(\mu_r X)$, (Figure 6). For both curves it will hold:

$$\frac{1/(\mu_i X)}{1/(\mu_r X)} = \eta \tag{70}$$

Rearranging Equation 70 we have

$$\frac{1}{\mu_r X} = \frac{1}{\mu_i X \eta} \tag{71}$$

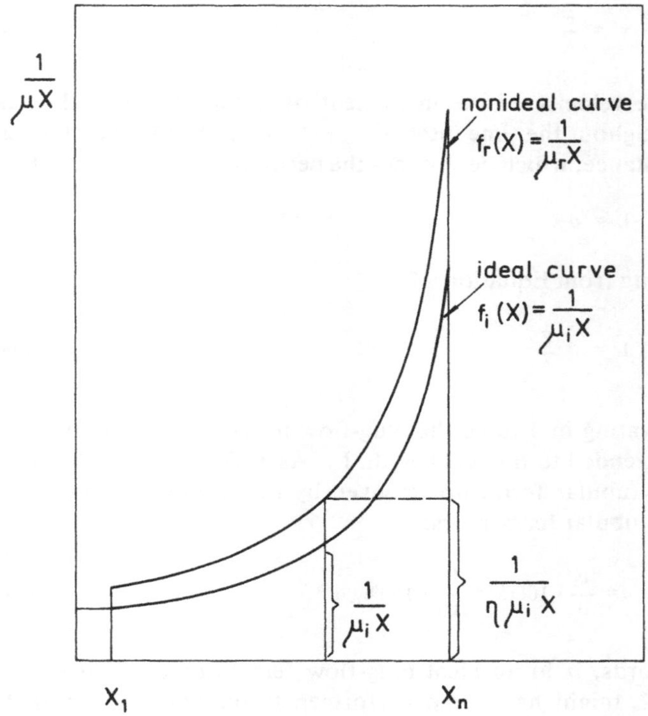

FIGURE 6. Difference in holding times between ideal and nonideal (real) tubular fermentors. Calculated for η = 40%. Explanation — see text.

from which it follows that in a nonideal situation for an arbitrary X, the reciprocal value increases (Figure 6), as the growth rate is reduced by η, $\mu_r X = \eta\mu_i X$. Thus, in order to achieve the concentration X_n, it is necessary to increase the holding time, θ_r, in the nonideal (real) tubular fermentor. This increase of the holding time, $\Delta\theta = \theta_r - \theta_i$ is represented by the area between both curves for the given entering, X_1, and leaving, X_n, concentrations:

$$\Delta\theta = \int_{X_1}^{X_n} [\, fr(X) - fi(X) \,]\, dX = \int_{X_1}^{X_n} \frac{1-\eta}{\eta} \frac{1}{\mu_i X}\, dX \qquad (72)$$

where the coefficient, η, is a complex value, in which also coefficients characterizing the effect of gas hold-up, axial mixing, settling of cells, and of other important factors are included. It should be specified in agreement with the design of the individual gradient fermentor.

B. Simulation of Plug-flow Fermentor

A continuous tubular plug-flow fermentor can be simulated by a series of continuous stirred fermentors having very short individual holding times. The total holding time, θ_T, is required for the increase of the concentration from X_1 up to X_n. For a given number of stages, N, a ratio exists $\theta_1 : \theta_2 : \ldots : \theta_N$, at which the total holding time, θ_T, represented by the sum of rectangle areas is minimum approaching the ideal curve $f_i(X)$, (Figure 7). This optimum ratio of holding times is dependent only on the shape

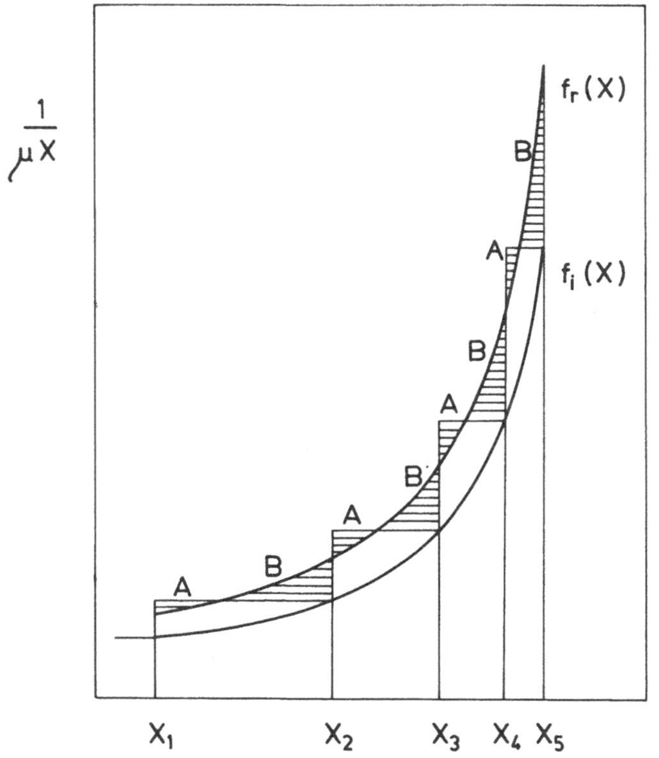

FIGURE 7. Simulation of a plug-flow fermentor by a multi-stage system. X — concentration of microorganisms in individual stages; $1/\mu X$ — reciprocal value of growth rate; A, B — hatched areas of triangles; $f_i(X)$, $f_r(X)$ — ideal and nonideal (real) curves; volumes or holding times in stages are identical, but differences between cell concentrations are progressively diminishing, $(X_2 - X_1) > (X_3 - X_2) > (X_4 - X_3) > (X_5 - X_4)$.

of the curve $f_i(X)$. The determination of this ratio is rather difficult and its calculation without use of a computer is possible only for a low number of stages, e.g., N = 2, 3, or 4.

Let us consider stages in series identical in volumes simulating the plug-flow fermentor. If the number of stages, N, is increased, the individual holding times or volumes are decreased. When proceeding in this way the total holding time, θ_T, necessary for achieving the needed concentration, X_n, decreases too, until the extreme case is reached, when the number of stages approaches infinitum, $N \rightarrow \infty$, and θ_T becomes equal to the holding time of the plug-flow fermentor (Figure 7). It means that it is possible to find such a number of stages, when the simulation of a nonideal tubular fermentor will be advantageous with respect to corresponding holding times. This situation will arise when the sum of triangle areas, A, above the curve $f_r(X)$ is lower than the sum of triangle areas, B, below the curve (Figure 7, hatched areas). In our example, in which the difference in the effectiveness between $f_i(X)$ and $f_r(X)$ amounts to 40%, four stages may sufficiently substitute for the nonideal plug-flow fermentor. If the same holding period is distributed to more than four stages (Figure 7), the holding time of the nonideal tubular fermentor becomes higher than the total holding time for that chosen N, $\theta_r > \theta_T$, and the following inequality will hold true for the advantageous number of stages:

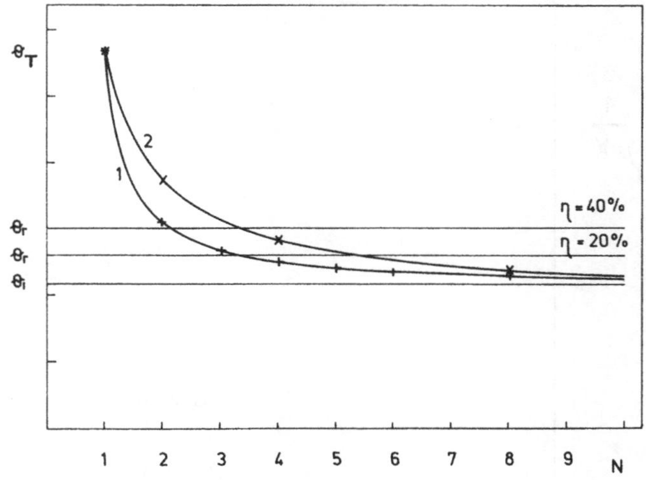

FIGURE 8. Changes in total holding time in dependence on the number of stages in series simulating the plug-flow fermentor. θ_T — total holding time of a series of stages; θ_i, θ_r — holding times of ideal and nonideal (real) plug-flow fermentors; N — number of stages (fermentors); 1 — different numbers of stages computerized from Equation 73, in which the same concentration of cells, X_n, is to be achieved; 2 — number of stages, designed for the same purpose, obtained by doubling of stages and reducing individual volumes (Figure 2).

$$\int_{X_1}^{X_N} f_r(X)dX > \sum_{n=1}^{N} (X_n - X_{n-1}) f_i(X_n) \qquad (73)$$

Since θ_T of a series composed of seven stages approaches already very near θ_i (Figure 8), a system consisting of the first stirred fermentor followed by a series of five up to seven stirred stages could be used. When designing such a multi-stage system it is usually found that individual volumes of stages in a series following the first fermentor are almost equal or smaller than one-third of the volume of the first fermentor.

In a series of fermentors situated horizontally and interconnected by a single short liquid stream, distinct from gas flow, the effect of slugging, fluid back-mixing and gas hold-up is reduced to a minimum. Culture conditions can be controlled separately in each stage.

In a tower, consisting of a series of stirred compartments separated from each other by a perforated plate, the effects of fluid back-mixing, slugging, gas hold-up and pressure drop can be significant, considerably reducing the operational effectiveness and the cultivation applicability. In such cases behavior of the tower could approach that of a unit thoroughly mixed, resembling more or less a stirred single-stage continuous fermentor.

When it is intended to design a tower fermentor, attention should be given to the construction of the inside arrangement of the tube and also to rheological properties of the culture, in order to minimize the unfavorable effects influencing the flow characteristics of the tower fermentor. As pointed out by Blanch and Bhavaraju,[50] knowledge of the flow behavior of non-Newtonian broths serves as a prerequisite for the understanding of the heat and mass transfer involved in the process design. Relatively little information is available concerning the fermentation broth rheology or the time changes in oxygen transfer rates, bubble formation, mixing effects, and of the agitation power requirement during the fermentation. The behavior of fermentation broths is of considerable importance from the point of view of cellular kinetics and in the engineering design of tower fermentors.

The example presented here was concerned with a part of the population growth period. However, the period of the product formation can be described in a similar way. Being simultaneous with the growth period, the separate solution of its spatial distribution is not necessary. That part of product accumulation, of course, which takes place over the growth period, is to be again distributed into further stages, thus increasing the number, but reducing volumes of fermentors in series.

IV. MULTI-STREAM SYSTEMS

If to a fermentor that is a member of a series of interconnected fermentors further inflows are introduced, multi-stream systems originate. They can be divided into two main groups based on the character of the added material: (1) the substrate, and (2) the population of microorganisms.

A. Inflow of Further Substrate

The substrate is fed in the form of a fresh culture medium. If other substances distinct from the substrate are used, they are fed dissolved in various solutions. The substrate added in a solid form can also be taken into account. Continuous cellulase production developed by Mitra and Wilke[51] may serve as example of such a system. In the first stage glucose is utilized and cellulose is decomposed in subsequent stages. A significant increase in the enzyme productivity was obtained in a multi-stage operation as compared with the single-stage fermentation. The pure spruce wood cellulose was fed by means of a volumetric disc feeder.

New inflows may be installed in an arbitrary n-th fermentor, but usually the fresh substrate is dosed to the second stage. All components present there, as well as those entering from the first stage are diluted by this further inflow, F_{02}, thus increasing the dilution rate, $D_2 = (F_1 + F_{02})/V_2$. The concentration of microorganisms, X_1, leaving the first stage is diluted to X_{02}:

$$X_{02} = \frac{F_1 X_1}{F_1 + F_{02}} \quad (74)$$

The substrate leaving the first stage, S_1, and the substrate added to the second stage are diluted to S_{02}:

$$S_{02} = \frac{F_1 S_1 + F_{02} S}{F_1 + F_{02}} \quad (75)$$

Balance equations for microorganisms and the substrate maintained under steady-state conditions may be written as follows:

$$D_2 X_{02} + \mu_2 X_2 = D_2 X_2 \quad (76)$$

$$D_2 S_{02} - \frac{\mu_2 X_2}{Y_2} = D_2 S_2 \quad (77)$$

If the product is already formed in the first stage, it is diluted in the same way and a similar equation may be written:

$$D_2P_{02} + p_2X_2 = D_2P_2 \quad (78)$$

In this system cells in a suitable physiological state prepared within the first stage enter the second stage, where they are shifted up by the dosed fresh medium. Since the second stage is continuously inoculated by the cells leaving the preceding stage, the inflow can be increased up to high values without a complete washing out of cells. The organisms are grown in excess substrate and conditions can be attained, under which the activity of enzymes and the specific rate of growth and product formation reach maximum values in the given medium. These maximum values have usually been found to be higher than those estimated in the batch culture, as shown by Řičica,[52] Řičica et al.,[53] and Fencl et al.,[54] who investigated enzymes involved in the primary metabolism. This phenomenon is of the same character as that called the hypertrophic growth by Powell.[55] As found by Miura et al.,[56] in the steady state the cellular content of the DNA and tRNA were constant regardless of μ, while the total RNA and rRNA contents were linearly dependent on μ. Thus, it can be postulated that in the excess substrate, due to a further inflow of the fresh medium, the cell division is limited and controlled by the rate of DNA synthesis and proceeds separately from the growth. On the other hand, the cell mass increases controlled by the increasing rate of rRNA synthesis and by the feed rate of rRNA for protein synthesis. This is manifested by cell prolongation, $\mu = k\,[R/X-(R/X)_c]$, where k is a constant, R/X is the actual concentration of rRNA, and $(R/X)_c$ is a minimum content of rRNA necessary for growth found in cells from the stationary phase of the batch.

B. Inflow of Further Microorganisms

If the culture leaving the single-stage or multi-stage system is concentrated by a continuous separation, and a fraction of that concentrate is recycled to a pertinent stage a biofeedback system is formed. Usually the concentrated culture is fed back into the first stage. Herbert[1] and Powell and Lowe[5] developed the basic theory, but their definitions of the back flow are different. In the theory of the former author, the concentrated cell suspension, CX, is returned at a flow rate α F, C termed as the concentration factor, $C \geqslant 1$, and α the volumetric feedback ratio, $\alpha \geqslant 0$, i.e, the ratio of feedback flow to final system output flow. The latter authors consider a fraction, a, of the emergent liquid volume and a fraction, b, of cells both being returned to the first fermentor. The definitions of Herbert[1] and Powell and Lowe[5] reciprocate in relations:

Feedback ratio:

$$\alpha = \frac{a}{1-a} \quad (79)$$

Concentration factor:

$$C = \frac{b}{a} \quad (80)$$

For the sake of a clearer arrangement it is advantageous to follow Herbert's[1] theory. If in a series of N stages, the cells are returned to an arbitrary n-th stage, we may write:

increase = input − output + growth + feedback

$$\frac{dX'_n}{dt} = D'_n(X'_{n-1} - X'_n) + \mu(S'_n)X'_n + \alpha CD'X'_N \quad (81)$$

increase = input − output − consumption + feedback

$$\frac{dS'_n}{dt} = D'_n(S'_{n-1} - S'_n) - \frac{\mu(S'_n)}{Y_n} X'_n + \alpha D'S'_N \tag{82}$$

where all values are marked with apostrophes.

The balance equations for X and S in other stages are identical with those developed for a single-stream system (Equations 7 and 26), except for equations for the stage with the feedback that are supplemented by the terms, $\alpha\ CD'X'_N$ and $\alpha\ D'S'_N$. Usually the concentrated cells are returned to the first stage.

In our further considerations we will use Equations 83 and 84 modified only for a single-stage system with the feedback, subscript N = 1, since conclusions drawn from that system are applicable to an arbitrary number of stages in series.

increase = input + feedback − output + growth

$$\frac{dX'_1}{dt} = 0 + \alpha D'CX'_1 - (1+\alpha)D'X'_1 + \mu(S'_1)X'_1 \tag{83}$$

increase = input + feedback − output − consumption

$$\frac{dS'_1}{dt} = D'S'_1 + \alpha DS'_1 - (1+\alpha)D'S'_1 - \frac{\mu(S'_1)}{Y_1} X'_1 \tag{84}$$

It follows from Equation 83 that the dilution rate in a system with feedback can be higher than in a system without feedback and that it can even exceed μ_{max}:

$$D' = \frac{\mu(S'_1)}{1 + \alpha - \alpha C} \tag{85}$$

Steady-state concentrations are defined by solving Equations 83 and 84:

$$\bar{S}'_1 = \frac{K_S A D'}{\mu_{max} - AD'} \tag{86}$$

$$\bar{X}'_1 = \frac{Y}{A}(S - S'_1) \tag{87}$$

$$\bar{X}'_E = Y(S - S'_1) \tag{88}$$

where $\overline{X}'_E$ is the concentration of cells in the final effluent. The coefficient, $A = 1 + \alpha - \alpha\ C = X'_E/X'_1$, acquires values within the range $0 < A \leqslant 1$. For, $A = 1$, the feedback is either zero, $\alpha = 0$, or the system works with a feedback, but without a separator concentrating cells, $C = 1$. The case for, $A = 0$, represents the limit situation (in fact not obtainable), when the total feedback of cells is reached. Final effluent concentration is zero, $X'_E = 0$, and concentration, X'_1, tends to infinity. Real values of α and C at different A are demonstrated in Figure 9.

Single-stage systems with (Equations 86, 87, and 88) and without the feedback (Equations 5 and 29) can be mutually compared, (Figure 10). The feedback operation enables the increase of dilution rate and even the critical one, D'_C, is significantly higher:

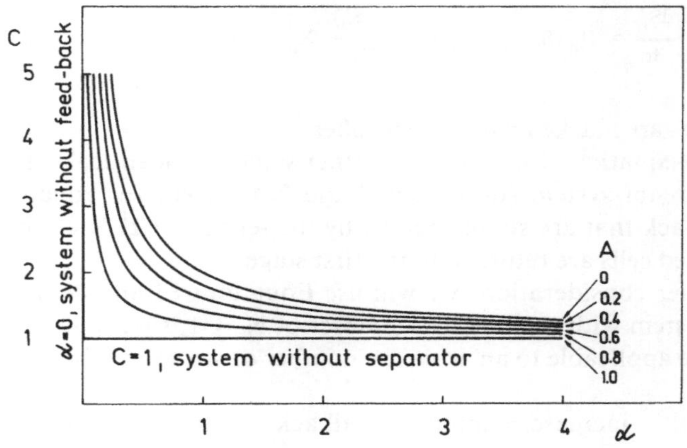

FIGURE 9. The range of possible values of the volumetric feedback ratio, α, and the concentration factor, C, at different coefficients, A, ($A = 1 + \alpha - \alpha C$).

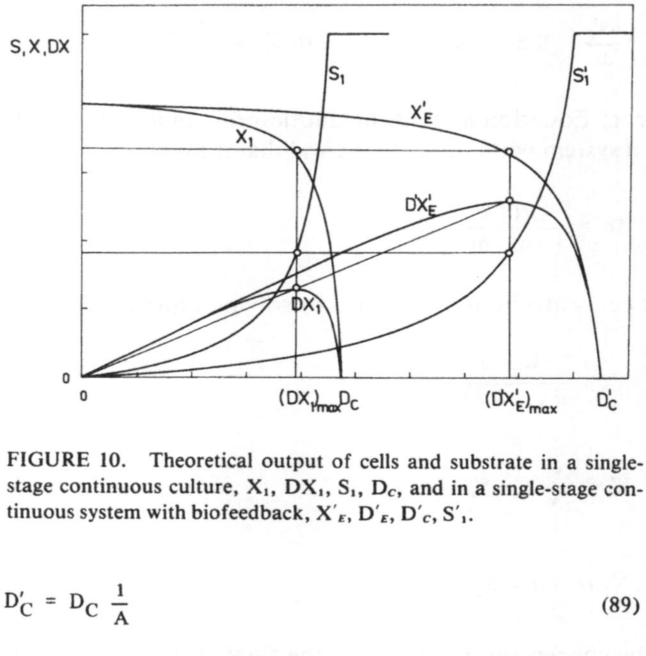

FIGURE 10. Theoretical output of cells and substrate in a single-stage continuous culture, X_1, DX_1, S_1, D_C, and in a single-stage continuous system with biofeedback, X'_E, D'_E, D'_C, S'_1.

$$D'_C = D_C \frac{1}{A} \tag{89}$$

Higher dilution rates provide a higher final output, $D'X'_E$, although the maximum attainable concentrations of cells leaving both systems are equal. This phenomenon can be best demonstrated by dilution rates providing the maximum output, $(D'\overline{X}'_E)_{max} > (D_1\overline{X}_1)_{max}$, where $\overline{X}'_E = \overline{X}_1$ and $\overline{S}'_E = \overline{S}_1$, (Figure 10).

If the comparison is made at the same and single dilution rate (in a vertical direction), it is apparent that the substrate concentration is lowered due to the feedback, $\overline{S}'_E < \overline{S}_1$ and the concentration of cells is increased, $\overline{X}'_E > \overline{X}_1$. In other words, if the metabolic activity of microorganisms is retained, the biofeedback can increase the effectiveness of substrate conversion.

A single-stage system with feedback attains the maximum output, $(D'X'_E)_{max}$, at a dilution rate, which conforms to the precondition:

$$\frac{d(D'X'_E)}{dD'} = 0 \tag{90}$$

By solving Equation 90, the optimum dilution rate is determined:

$$D'_{opt} = \frac{\mu_{max}}{A}\left(1-\sqrt{\frac{K_S}{K_S + S}}\right) \tag{91}$$

If Equations 86, 87, and 88 are modified using Equation 91, the steady-state concentrations of cells, X'_E, and of the substrate, S'_E at D'_{opt} can be derived. These terms are identical with Equations 35 and 36 for a single-stage without feedback.

When comparing maximum outputs in systems with the feedback and without it both the dilution rates and outputs are related as follows:

$$D' = D\frac{1}{A} \tag{92}$$

$$(D'\bar{X}'_E)_{max} = (D_1\bar{X}_1)_{max}\frac{1}{A} \tag{93}$$

Culture systems with the biofeedback are used mostly in treatment of waste waters. Let us mention here some papers concerning this problem: Ramanathan and Gaudy,[57,58] Srinivasaraghavan and Gaudy,[59] Gaudy et al.,[60] Sundstrom et al.,[61] Bonotan-Dura and Yang,[62] and Wong and Yang.[63] Some critical views on tanks with piston-flow feed or tanks with complex sewage mixing with partial recirculation of the activated sludge were presented by Schmidt-Holthausen.[64] A theoretical model for a continuous multi-stage enzyme production system, which consists of a growth fermentor with glucose feeding without the enzyme induction and a subsequent series of well-stirred vessels with a recycle, in which cellulose solids are contacted with organisms in order to induce the enzyme formation, was developed by Von Stockar et al.[65]

A pertinent case is to be mentioned here, when a culture of microorganisms is cultivated separately and fed as a substrate to a culture of another organism. Lysogenic cells grown in a single-stage system might be dosed to a further fermentor in which an induced phase population develops. In another prey-predator or host-parasite system, bacteria might be used for feeding a slime mold, or protozoa. In prey-predator systems described in the literature, both organisms are usually grown together in a common single-stage system. This system behaves as a special type of a mixed population and, hence, is out of the scope of this chapter. Mixed populations, however, are treated in one of the following chapters.

A multi-stage branched system, termed "ecostat", for exploring ecological interactions between a number of bacterial species was presented by Parker.[66] In the procedure, steady-state populations of different species are added simultaneously to a common mixed culture vessel. Although the medium in the common vessel probably represents a type of a chemostatic medium, in which one of the nutrients must be first exhausted, there are apparently differences between the "multifactor" limitation arising from the greatly reduced concentrations of all factors in this system and a single factor starvation in true chemostatic media. An equilibrium established in a mixed population can be changed by changing character of the medium or the infeed rate of pure cultures.

In order to eliminate the loss of the ability to produce antibiotics, Reusser[67] and Ping Shu[68] used a cross-flow culture system.There are two first fermentors in parallel that are connected with a successive multi-stage system. One of the two fermentors runs continuously for a predetermined period of several days. Before the culture loses its production capacity, a new batch culture is prepared in the other fermentor and grown until a certain state of culture development is attained. Then the supply of fresh medium is started and the batch culture is changed to a continuous one. The fermentor operating continuously is stopped, the antibiotic production is completed under batch conditions, and the product is isolated. The alternation of both the first fermentors functioning as the first stage is repeated several times.

C. Dialysis Fermentation

A dialysis culture of microorganisms can also be classified as a complex system. Schultz and Gerhardt[69] summarized the design, the theory, and the application of different types of dialysis fermentation. The growth of microorganisms or cells of higher organisms is carried out in a culture vessel that is remote from, but in communication with a nutrient reservoir vessel. All three principal regions — culture, dialysis, and reservoir — are separated. This separation makes it possible for each to be controlled independently and effectively. In a dialysis culture system the culture vessel and the nutrient reservoir both can be of conventional design without any special modifications. A suitably designed dialyzer is connected with the culture vessel and with the reservoir by tubing and pumps. There are four modes of operating a dialysis culture, however, for a continuous cultivation the fermentor is operated in steady-state with constant inflow of fresh medium and outflow of the culture. The population is permanently recirculated through the dialyzer. The same medium can either flow through the dialyzer and directly out or be recycled back to a reservoir being operated in steady-state. Note that the other specific substrate can be added by dialysis, too.

Since the theory and detailed mathematical description of all types of dialysis culture were developed by Schultz and Gerhardt,[69] let us mention here only few equations describing the growth of microorganisms in the culture region, presupposing that no cells are present in the fermentor feed stream. The growth in steady-state may be described by Equation 5. The cells utilize the limiting substrate by the specific consumption rate:

$$qX = -\frac{\mu X}{Y} - Y_E X \tag{94}$$

where Y_E is "specific maintenance rate" defined by Marr et al.,[70] and $Y_E X$ represents the portion of the substrate that is used by the population to maintain itself. The steady-state balance equation for the substrate in the culture region, V, may be written:

$$\text{input} - \text{consumption} = \text{output}$$

$$DS + \frac{P_m A_m (S_D - S_V)}{V} - qX = DS_V \tag{95}$$

where, S, and S_V, are concentrations of the substrate entering and leaving the culture region; S_D, is the substrate concentration in the reservoir medium leaving the dialyzer; P_m, is the permeability coefficient, $[m \cdot t^{-1}]$, which must be experimentally determined for each membrane and solute; A_m, is the total area of the membrane, $[m^2]$, and $P_n A_m(S_D - S_V)$, represents the removal of the substrate by dialysis.

Similarly the steady-state balance equation for the substrate concentration in the dialysis chamber is:

$$\text{input} = \text{removal} + \text{output}$$

$$F_D S_R + P_m A_m S_V = P_m A_m S_D + F_D S_D \tag{96}$$

$$S_D = \frac{F_D S_R + P_m A_m S_V}{P_m A_m + F_D} \tag{97}$$

where, S_R is the concentration of the substrate in the reservoir medium entering the dialyzer by the volumetric rate, F_D.

The most striking and important difference between dialysis and nondialysis culture is the much higher cell density attainable in dialysis culture, especially at low dilution rates, due to an additional supply of the substrate through the dialysis membrane and to removal of inhibiting metabolites from the culture. The concentration of cells can be derived from Equation 95:

$$X = \frac{F(S - S_V) + P_m A_m (S_D - S_V)}{qV} \tag{98}$$

The degree of complexity increases when the formation of a particular microbial product in relation to dialysis culture is considered. However, if a kinetic model of a product formation in nondialysis culture is available, it is not too difficult to predict the behavior in a dialysis culture. The only additional information needed is the permeability characteristics of the product for the dialysis membrane material, P'_m. If the product is intracellular or nondialyzable, $P'_m = 0$, the culture region behaves like a single-stage continuous fermentor. If the product is permeable through the membrane, its concentration is permanently lowered by dialysis and for steady-state conditions the balance equation may be written similarly with Equation 47:

$$pX = \frac{P'_m A_m (P_D - P_V)}{V} + DP_V \tag{99}$$

where, P_V, is a product concentration in the culture region and, P_D, in the reservoir medium leaving the dialysis chamber.

The main use of dialysis culture has been in the production of cell mass and macromolecular products such as enzymes and toxins, antigens, allergens, polymers, and even viruses and cell particles. The obvious application is in the manufacture of vaccines. In principle, the dialysis technique also allows the production of diffusible products inhibitory for the producer, as the concentration can be maintained below the critical level by means of dialysis.

D. Solid Substrate Fermentation

In recent years, the interest in feeding animals fodder mixtures prepared by fermentation of solid substrates containing cellulose has come forward. In order to eliminate the rise of further excessive waste waters the ground material is moistened only to a certain degree of humidity by aqueous solutes of inorganic phosphorus and nitrogen salts. For the same purpose other wastes containing phosphorus and nitrogen are also used. The mixture is then fermented by cellulolytic microorganisms. Better decomposition of the material by microorganisms and sterility can be enabled by pretreatment of the mixture by chemical hydrolysis under increased pressure and temperature.

The culture device designed for solid substrate fermentation can be arranged in vertical or horizontal position. The vertical type is composed from chambers positioned each above the other forming a towerlike system. With the horizontal type a tubularlike fermentor is formed. The first stage of both systems into which the substrate and solutes are separately fed, is thoroughly mixed. There the period of maximum vegetative growth of organisms decomposing the cellulose is taking place. The dosing of the solid substrate and of solutes is conditioned by the rate of cellulose decomposition. The partly fermented mass, being kept all the time loose or up to a mash consistence, is then transported to successive stages.

In a tower system the fermented material inside the chambers is aerated by the upward airflow and more or less mechanically agitated and forced to fall from one chamber through the bottom down into the next one. In a horizontal tubularlike culture design, the fermented mass is transported along the tube with the aid of a spiral or a running belt, being mechanically loosened and aerated along the way. The holding time in both systems is determined by the needed degree of cellulose decomposition. The microorganisms decomposing cellulose can be used as a pure culture or as a mixture of different cellulolytic organisms.

REFERENCES

1. **Herbert, D.,** A theoretical analysis of continuous culture systems, in *Continuous Culture of Microorganisms,* S. C. I. Monograph No. 12, Soc. Chem. Ind., London, 21, 1961.
2. **Herbert, D.,** Multi-stage continuous culture, in *Continuous Cultivation of Microorganisms,* Proc. 2nd Symp. Prague, June 18—23, 1962, Målek, I., Beran, K., and Hospodka, J., Eds., Publishing House Czechoslovak Academy of Science, Prague, 1964, 23.
3. **Aiba, S., Humphrey, A. E., and Millis, N. F.,** *Biochemical Engineering,* University of Tokyo Press, Tokyo, 1965, chap. 4 and 5.
4. **Fencl, Z.,** Theoretical analysis of continuous culture systems, in *Theoretical and Methodological basis of Continuous Culture of Microorganisms,* Målek, I. and Fencl, Z., Eds., Publishing House Czechoslovak Academy of Science, Prague, 1966, chap. 3.
5. **Powell, E. O. and Lowe, J. R.,** Theory of multi-stage continuous cultures, in *Continuous Cultivation of Microorganisms,* Proc. 2nd Symp. Prague, June 18—23, 1962, Målek, I., Beran, K., and Hospodka, J., Eds., Publishing House Czechoslovak Academy of Science, Prague, 1964, 45.
6. **Bischoff, K. B.,** Optimal continuous fermentation reactor design, *Can. J. Chem. Eng.,* 44, 281, 1966.
7. **Aris, R.,** *The Optimal Design of Chemical Reactors. A Study in Dynamic Programming,* Academic Press, New York, 1961, chap. 3 and 5.
8. **Denbigh, K. G., and Turner, J. C. R.,** *Chemical Reactor Theory. An Introduction,* 2nd ed., Cambridge University Press, London, chap. 3—6.
9. **Kafarov, V.,** *Cybernetic Methods in Chemistry and Chemical Engineering,* Mir Publishers, Moscow, 1976, chap. 2 and 4.
10. **Luedeking, R. and Piret, E. L.,** Transient and steady states in continuous fermentation. Theory and experiment, *J. Biochem. Microbiol. Technol. Eng.,* 1, 431, 1959.
11. **Kono, T.,** Kinetics of microbial cell growth, *Biotechnol. Bioeng.,* 10, 105, 1968.
12. **Kono, T. and Asai, T.,** Kinetics of continuous cultivation, *Biotechnol. Bioeng.,* 11, 19, 1969.
13. **Kono, T. and Asai, T.,** Kinetics of fermentation processes, *Biotechnol. Bioeng.,* 11, 293, 1969.
14. **Powell, E. O.,** The growth rate of microorganisms as a function of substrate concentration, in Microbial Physiology and Continuous Culture, 3rd Int. Symp. M. R. E., Porton Down, Powell, E. O., Evans, C. G. T., Strange, R. E., and Tempest, D. W., Eds., Her Majesty's Stationary Office, 1967, 34.
15. **Powell, E. O.,** Transient changes in the growth rate of microorganisms, in *Continuous Cultivation of Microorganisms,* Proc. 4th Symp. Prague, June 17—21, 1968, Målek, I., Beran, K., Fencl, Z., Munk, V., Řičica, J., and Smrčkovå, H., Eds., Academia, Prague, 1969, 275.
16. **Schulze, K. L. and Lipe, R. S.,** Relationship between substrate concentration, growth rate, and respiration rate of *E. Coli* in continuous culture, *Arch. Microbiol.,* 48, 1, 1964.
17. **Contois, D. E.,** Kinetics of bacterial growth: relationship between population density and specific growth rate of continuous cultures, *J. Gen. Microbiol.,* 21, 40, 1959.

18. **Pirt, S. J.**, The maintenance energy bacteria in growing culture, *Proc. R. Soc. London, Ser. B*, 163, 224, 1965.
19. **Van Uden, N.**, Transport-limited growth in the chemostat and its competitive inhibition; a theoretical treatment, *Arch. Microbiol.*, 58, 145, 1967.
20. **Van Uden, N.**, Kinetics of nutrient-limited growth. *Annu. Rev. Microbiol.*, 23, 473, 1969.
21. **Edwards, V. E. and Wilke, C. R.**, Analytical Methods in Bacterial Kinetics, Lawrence Radiation Laboratory, University of California, Berkeley, (AEC Contract No. W-7405-eng-48), 1967.
22. **Aiba, S., Nagai, S., Endo, I., and Nishizawa, Y.**, Dynamical analysis of microbial growth, *AICHE J.*, 15, 624, 1969.
23. **Nikolayev, P. J. and Sokolov, D. P.**, Evaluation of coefficients of equations describing process of the microbial cultivation, (in Russian), *Prikl. Biokhim. Mikrobiol.*, 4, 562, 1968.
24. **Dabes, J. N., Finn, R. K., and Wilke, C. R.**, Equations of substrate limited growth: the case for Blackman kinetics, *Biotechnol. Bioeng.*, 15, 1159, 1973.
25. **Nyholm, N.**, Kinetics of phosphate limited algal growth, *Biotechnol. Bioeng.*, 19, 467, 1977.
26. **Bijkerk, A. H. E. and Hall, R. J.**, A mechanistic model of the aerobic growth of *Saccharomyces cerevisiae*, *Biotechnol. Bioeng.*, 19, 267, 1977.
27. **Keller, A. K. and Gerhardt, P.**, Continuous lactic acid fermentation of whey to produce a ruminant feed supplement high in crude protein, *Biotechnol. Bioeng.*, 17, 997, 1975.
28. **Herbert, D.**, Stoicheiometric aspects of microbial growth, in *Continuous Culture 6: Applications and New Fields*, Dean, A. C. R., Ellwood, D. C., Evans, C. G. T., and Melling, J., Eds., Ellis Horwood, Ltd., Chichester, 1976, 1.
29. **Van Dedem, G. and Moo-Young, M.**, A model for diauxic growth, *Biotechnol. Bioeng.*, 17, 1301, 1975.
30. **Tsao, G. T. and Yang, C. M.**, Extended Monod equation, *Biotechnol. Bioeng.*, 18, 1827, 1976.
31. **Yoon, H., Klinzing, G., and Blanch, H. W.**, Competitition for mixed substrates by microbial populations, *Biotechnol. Bioeng.*, 19, 1193, 1977.
32. **Sinclair, C. G. and Ryder, D. N.**, Models for the continuous culture of microorganisms under both oxygen and carbon limiting conditions, *Biotechnol. Bioeng.*, 17, 375, 1975.
33. **Cooney, C. L. and Wang, D. I. C.**, Transient response of *Enterobacter aerogenes* under a dual nutrient limitation in a chemostat, *Biotechnol. Bioeng.*, 18, 189, 1976.
34. **Řičica, J.**, Recent theoretical and practical trends in continuous cultivation, in *Fermentation Advances*, Perlman, D., Ed., Academic Press, New York, 1969, 427.
35. **Johnson, D. B. and Berthouex, P. M.**, Efficient biokinetic experimental designs, *Biotechnol. Bioeng.*, 17, 557, 1975.
36. **Johnson, D. B. and Berthouex, P. M.**, Using multiresponse data to estimate biokinetic parameters, *Biotechnol. Bioeng.*, 17, 571, 1975.
37. **Williamson, K. J. and McCarty, P. L.**, Rapid measurement of Monod half-velocity coefficients for bacterial kinetics, *Biotechnol. Bioeng.*, 17, 915, 1975.
38. **Dean, A. C. R. and Hinshelwood, C.**, *Growth Function and Regulation in Bacterial Cells*, Clarendon Press, Oxford, 1966, chap. 6.
39. **Dean, A. C. R., Ellwood, D. C., Melling, J., and Robinson, A.**, The action of antibacterial agents on bacteria grown in continuous culture, in *Continuous Culture 6: Applications and New Fields*, Dean, A. C. R., Ellwood, D. C., Evans, C. G. T., and Melling, J., Eds., Ellis Horwood, Ltd., Chichester, 1976, 251.
40. **Aiba, S., Shoda, M., and Nagatani, M.**, Kinetics of product inhibition in alcohol fermentation, *Biotechnol. Bioeng.*, 10, 845, 1968.
41. **Ierusalimski, N. D.**, Bottle-necks in metabolism as growth rate controlling factors, in Microbial Physiology and continuous culture, 3rd Int. Symp. M. R. E., Porton Down, Powell, E. D., Evans, C. G. T., Strange, R. E., and Tempest, D. W., Eds., Her Majesty's Stationery Office, 1967, 23.
42. **Driessen, F. M., Ubbels, J., and Stadhouders, J.**, Continuous manufacture of yogurt. I. Optimal conditions and kinetics of the prefermentation process, *Biotechnol. Bioeng.*, 19, 821, 1977.
43. **Driessen, F. M., Ubbels, J., and Stadhouders, J.**, Continuous manufacture of yogurt. II. Procedure and apparatus for continuous coagulation, *Biotechnol. Bioeng.*, 19, 841, 1977.
44. **Wick, E. and Popper, K.**, Continuous fermentation in slant tubes, *Biotechnol. Bioeng.*, 19, 235, 1977.
45. **Greenshields, R. N. and Smith, E. L.**, The tubular reactor in fermentation, *Biochemistry*, 9(3), 11, 1974.
46. **Štěrbáček, Z.**, Design methods for tubular fermentation systems, *Folia Microbiol. (Prague)*, 20, 171, 1975.
47. **Atkinson, B. and Knights, A. J.**, Microbial film fermenters: their present and future applications, *Biotechnol. Bioeng.*, 17, 1245, 1975.

48. **Howell, J. A. and Atkinson, B.**, Influence of oxygen and substrate concentrations on the ideal film thickness and the maximum overall substrate uptake rate in microbial film fermenters, *Biotechnol. Bioeng.*, 18, 15, 1976.
49. **Moser, A.**, Dünnschichtreaktoren in der Biotechnologie, *Chem. Ing. Tech.*, 49, 612, 1977.
50. **Blanch, H. W. and Bhavaraju, S. M.**, Non-Newtonian fermentation broths: rheology and mass transfer, *Biotechnol. Bioeng.*, 18, 745, 1976.
51. **Mitra, G. and Wilke, C. R.**, Continuous cellulase production, *Biotechnol. Bioeng.*, 17, 1, 1975.
52. **Řičica, J.**, Experimental use of the two-stage pluristream continuous system, *Mikrobiologija*, 1, 155, 1964.
53. **Řičica, J., Nečinová, S., Stejskalová, E., and Fencl, Z.**, Properties of microorganisms grown in excess of the substrate at different dilution rates in continuous multistream culture systems, in Microbial Physiology and Continuous Culture, 3rd Int. Symp. M. R. E., Porton Down, Powell, E. O., Evans, C. G. T., Strange, R. E., and Tempest, D. W., Eds., Her Majesty's Stationary Office, 1967, 196.
54. **Fencl, Z., Řičica, J., Munk, V., and Novák, M.**, Physiological changes in filamentous organisms as a function of growth rate, in Microbial Physiology and Continuous Culture, 3rd Int. Symp. M. R. E., Porton Down, Powell, E. O., Evans, C. G. T., Strange, R. E., and Tempest, D. W., Eds., Her Majesty's Stationary Office, 1967, 186.
55. **Powell, E. O.**, Hypertrophic growth, *J. Appl. Chem. Biotechnol.*, 22, 71, 1972.
56. **Miura, Y., Tsuchiya, K., Nishikawa, K., Obata, T., and Okazaki, M.**, Behaviour of cell structural components in steady and transient states of growth of *Bacillus subtilis*, *J. Ferm. Technol.*, 52, 100, 1974.
57. **Ramanathan, M. and Gaudy, A. F., Jr.**, Effect of high substrate concentration and cell feedback on kinetic behaviour of heterogeneous populations in completely mixed systems, *Biotechnol. Bioeng.*, 11, 207, 1969.
58. **Ramanathan, M. and Gaudy, A. F., Jr.**, Steady-state model for activated sludge with constant recycle sludge concentration, *Biotechnol. Bioeng.*, 13, 125, 1971.
59. **Srinivasaraghavan, R. and Gaudy, A. F., Jr.**, Operational performance of an activated sludge process with constant sludge feedback, *J. Water Pollut. Control Fed.*, 47, 1946, 1975.
60. **Gaudy, A. F., Jr., Srinivasaraghavan, R., and Saleh, M.**, Conceptual model for design and operation of activated sludge processes, *J. Environ. Eng. Div. Am. Soc. Civ. Eng.*, 103, 71, 1977.
61. **Sundstrom, D. W., Klei, H. E., and Brookman, G. T.**, Response of biological reactors to sinusoidal variations of substrate concentration, *Biotechnol. Bioeng.*, 18, 1, 1976.
62. **Bonotan-Dura, F. M. and Yang, P. Y.**, The application of constant recycle solids concentration in activated sludge process, *Biotechnol. Bioeng.*, 18, 145, 1976.
63. **Wong, Y. K. and Yang, P. Y.**, Effects of quantitative shock loadings on the constant recycle sludge concentration activated-sludge process, *Biotechnol. Bioeng.*, 19, 43, 1977.
64. **Schmidt-Holthausen, H. J.**, Some critical views of selected processes for the biological treatment of sewage, *Process Biochem.*, 12(2), 27, 1976.
65. **Von Stockar, U., Yang, R. D., and Wilke, C. R.**, Computation of the fraction of induced cells in enzyme induction systems, *Biotechnol. Bioeng.*, 19, 445, 1977.
66. **Parker, R. B.**, Continuous-culture system for ecological studies of microorganisms, *Biotechnol. Bioeng.*, 8, 473, 1966.
67. **Reusser, F.**, Continuous fermentation of novobiocin, *Appl. Microbiol.*, 9, 366, 1961.
68. **Ping Shu**, Development of a cross-flow fermentation process with special reference to chlortetracycline production, *Biotechnol. Bioeng.*, 8, 353, 1966.
69. **Schultz, J. S. and Gerhardt, P.**, Dialysis culture of microorganisms: design, theory, and results, *Bacteriol. Rev.*, 33, 1, 1969.
70. **Marr, A. G., Nilson, E. H., and Clark, D. J.**, Maintenance requirement of *E. coli*, *Ann. N.Y. Acad. Sci.*, 102, 536, 1963.

Chapter 5

CONTINUOUS CULTURE IN THE FERMENTATION INDUSTRY*

V. R. Srinivasan and R. J. Summers

TABLE OF CONTENTS

I. Introduction 98

II. Processes Available for Development 100
- A. Production of Growth Associated Products 100
- B. Production of Non-Growth Associated Products 103
- C. Production of Transformation Products 106

III. Continuous Beer Production 107

IV. Far Range Development 108

V. Summary 109

References 110

* This chapter was submitted in June 1978.

I. INTRODUCTION

The application of continuous cultivation of microorganisms is not a novel concept. Initial attempts were made in the development of such processes at the beginning of this century; however, the results were unsatisfactory. Throughout this century a collective understanding of microbial physiology and biochemical regulation, and the theory of continuous cultivation together with technological advancements have led to the development of continuous fermentation having tremendous potentiality (Dawson[14]). The setbacks commonly encountered have been overcome at the bench-scale level, yet industrial development of pilot plant and full-scale continuous production processes has been proceeding at a relatively slow rate. The literature is scant with examples of pilot plant studies being conducted currently in the western world. It is not known whether this is due to a lack of development, or simply because of the proprietary nature of industrial practices. Few known examples of major importance are currently employed in the U.S. which may be due to the success of well established, more conventional batch fermentation processes. However, the depletion of natural resources and increased population pressure should provide the economic incentive in the near future for a major industrial commitment towards the development of continuous fermentations for microbial products.

Many advantages are associated with single and multistage continuous fermentations that make these processes desirable over conventional batch fermentations. The four principal advantages listed below will be discussed briefly, but as well summarized by Hospodka.([29])

1. Increased productivity
2. Uniformity of operation
3. Ease in automation
4. Prevention of contamination

Batch fermentation normally results in a typical growth curve where a high productivity is achieved only for a brief period during the growth cycle of the microbial population. The duration of the range of high productivity in batch culture is dependent upon a variety of factors including the specific fermentation product being produced, the medium involved, and the specific growth characteristics of the organism. However, continuous processes operate at maximum concentration of microbial cells over extended periods of time (several weeks to months). Specially designed fermenation apparatus and optimum conditions are employed to maximize product formation. In addition, some fermentation products are only produced during a very narrow range of a specific physiological state. The employment of the continuous process to maintain the population at such a physiological state for long periods of time increases productivity tremendously. Therefore, the overall productivity is significantly increased over that which is obtained from periodic fermentations of equal volume. Usually, the reactor volume can then be reduced — resulting in lower costs associated with fermentation design, construction, maintenance, and the use of raw materials for nutrients. This results in a decrease in expenses while productivity remains optimum. This is one of the main principal advantages associated with continuous fermentations and often shadows the others.

Although productivity is the primary concern of all industrial fermentations, the continuous process offers economical advantages associated with uniformity of operation. Batch fermentation requires peak work loads during fermentation preparation and product isolation and purification. During the actual fermentation little effort is required from a labor standpoint. Peak work loads are reduced during a continuous fermentation and there is little variation. Once the microbial population has been established the rates of nutrient usage, aeration, pH adjustment, cooling, and product formation remain constant for the entire length of most continuous fermentations. Continuous fermentation tends to provide simplicity of operation, and requires that all other processes supporting the fermentation be conducted on a continuous level also. Product isolation and purification are normally performed by methods that are continuous or easily adapted to continuous.

The uniformity of operation of continuous processes suggests relative ease in the automation of such a process. Batch fermentation occurs as an everchanging interaction between the external environment of the medium and an increasing microbial population. The growth response of the population is constantly being influenced by this interaction. Placing such a system under automative control would require an exact understanding of the interaction, including an elaborate monitoring system of all factors that would influence the fermentation process. Such a system is feasible only during fermentations where the external environment is clearly defined and optimized for the concentration of the microbial population. Unfortunately the majority of fermentation processes have not been examined in such a manner and automative control would be difficult. The continuous process, on the other hand, is a self-regulating system. During the period of maximum productivity the growth rate of the population is equal to the rate of entry of the growth-limiting nutrient. When this has been achieved, the population is said to exist in steady state and very little variation occurs in either the concentration of nutrients in the medium or the concentration of the population for the duration of the fermentation. Automative control of the parameters that do change is relatively simple and can be accomplished continuously in step with the fermentation. Usually maintenance of a constant pH, temperature, rate of aeration, and rate of cooling will suffice for insuring optimum conditions for the fermentation. In addition, continuous rapid assay techniques can be used to monitor the fermentation for the entire duration as a quality control measure. Any deviation from known steady-state values can be used to indicate an immediate problem occurring during fermentation, and appropriate and immediate corrective action can be taken.

Both the batch fermentation and the continuous fermentation have the same probability of contamination at the start of the fermentation cycle. Contamination usually occurs during inoculation, where the inoculum size is relatively small and the nutrient concentration is in excess. The ability of a contaminant to become established at this time is favored in both types of fermentation. By the same time the continuous process nears steady state, the batch fermentation usually is near completion. As the length of the fermentation increases there is also an increased chance for contaminating microoranisms to gain entry into the system. It should be kept in mind, the circumstances present at steady-state populations are not conducive for the establishment of an invading microorganism. The contaminant must not only grow in the medium, but must divide at a rate more rapid than the dilution rate of the culture. In addition, the nutrient concentration at steady state is very low, and one nutrient limits either the growth rate of the population or the population density. Therefore, the invading microorganisms must be able to compete for the limiting nutrient more efficiently than the already established microbial population. If the contaminating microorganism cannot overcome both of these obstacles it will eventually wash out the reactor vessel without becoming established. There are instances where slow growing microorganisms can become established if they possess strong adhesive properties and are not evenly dispersed in the liquid medium, as in the case of some fungi. Yet, when a heterogeneous population does exist that is not antagonistic, often times a new steady-state value is reached. Thus, it may be possible to continue the fermentation for the duration of the period at a lower level of productivity.

The advantages of continuous culture operations suggest an ideal system to conduct any type of biological fermentation. This is not necessarily true, and the selection of a suitable fermentation system requires an extremely thorough economic evaluation. Hospodka[29] points out that continuous processes currently in use on an industrial level are those which require inexpensive raw materials to produce very large amounts of an inexpensive product. Since the productivity is very high the economic advantage is similarly high. This chapter will serve as a review of the traditional industrial continuous fermentation processes and more importantly, indicate processes demonstrating potential for future development.

II. PROCESSES AVAILABLE FOR DEVELOPMENT

A. Production of Growth Associated Products

Continuous cultivation has served as an excellent tool for the investigation of growth-related problems and metabolite formation at the bench level. The early work was concerned with describing population variation to alterations of the external cellular environment. Within the last decade, continuous cultivation has been successfully used to examine specific intracellular biochemical events subject to sensitive environmental alterations. It is not unexpected, therefore, that the continuous production of growth associated end products was attempted early in the developmental years. The prewar and war years of the 1940s provided an impetus for examining inexpensive crude substrates for batch and continuous production of 2,3-butylene glycol,[23,48,75] alcohol,[4,8,61] and lactic acid.[58,59,76] Following the war years, economic circumstances led to a loss of interest in the full scale development of continuous biological production. These processes, however, have set the foundation for demonstrating the potentiality of applied continuous cultivation with undefined crude substrates.

The 2,3-butylene glycol fermentations of the the 1940s were typical batch processes using *Klebsiella aerogenes* (formerly *Aerobacter*). Perlman[48] examined wood hydrolysates as a source of carbon and energy for the production of 2,3-butylene glycol. Following acid hydrolysis of the wood chips the substrate is rich in hexoses, pentoses, and acetic acid. Urea, potassium phosphate, and calcium carbonate were the only additives following adjustment to pH 6. Results indicated variability in fermentability, which was dependent upon the source of the hydrolysate. An alkaline pretreatment of some hydrolysates was also accomplished, and in general, the alkali pretreatment reduced the fermentation time 25% with 90% utilization of reducing sugars within 48 hr. Perhaps this was due to the removal of inhibitory materials present in the hydrolysate. The product yield was slightly greater than 33%. Attempts were made to increase product formation by increasing sugar concentration. These studies resulted in a similar yield; however, the fermentation time also increased. Comparable results were obtained from acid-hydrolyzed starch substrates, however, the fermentation demonstrated a sensitivity to trace metal concentrations.[75] Fermentation media made up with distilled water produced a near theoretical yield of 2,3-butylene glycol with small amounts of acetoin and ethanol. The same medium made with tap water reduced the fermentation time to 24 hr, but the yield of the primary fermentation product was significantly reduced while the ethyl alcohol concentration increased. Further investigation indicated 2,3-butylene glycol formation was depressed by increases in the concentrations of manganese, cobalt, or molybdenum. Zinc or copper was found to stimulate primary product formation in some samples. Sensitivities such as these to trace metal involvement render an industrial fermentation difficult to standardize when using inexpensive, poorly defined nutrient sources.

The use of molasses substrates for glycol formation is advantageous due to the high sugar concentrations of the substrates. Freeman and Morrison[23] demonstrated similar fermentation times and yields were possible from molasses, but the amount of products formed was greater. Variability was also noted between the substrate sources, and the fermentation times between sucrose and molasses. This may be due to materials present in the crude substrates that interfere with the fermentation rate. Press liquor from citrus waste is another inexpensive molasses type substrate available for 2,3-butylene glycol production. The sugar content of press liquor is between 7 and 10% which is lower than most conventional molasses. Batch fermentation of the liquor produces lower yields; therefore, this substrate may not be as desirable as others previously mentioned. High glycol yields were obtained following concentration of the press liquor.[40]

Pirt and Callow[49] examined 2,3-butylene glycol formation by *K. aerogenes* in batch and continuous culture with sucrose in a defined medium. Total aerobic growth in batch culture produced no glycol, however, 15 g/ℓ 2,3-butylene glycol was produced from 100 g/ℓ sucrose under partially aerobic conditions. The acetoin concentration was also influenced by oxygen concentration and was reduced when only low levels of oxygen were provided. The rate of glycol production was three times greater in continuous culture (4.6g/ℓ/ hr) than batch culture (1.6 g/ℓ/ hr). It was interesting to note that glycol concentration and yield was far greater in batch cultures rather than continuous cultures. This was probably due to the nature of steady-state conditions, which requires an actively growing population and is constantly removing a portion of the population. Thus, the optimum fermentation for glycol production would be a two-stage process. The first stage should be designed for rapid growth, and the second stage for product formation where the bacteria are retained in the vessel.

Alcohol production by microbial fermentation has been of interest for centuries, and continuous fermentation by distilleries has been attempted since as early as 1915. Early advances in continuous alcohol production from carbohydrates were accomplished by Russian investigators and their work is well summarized by Hospodka.[29] Two types of alcohol are produced by industrial fermentation processes. Industrial alcohol is usually produced from molasses, whereas beverage alcohol is principally produced from grains and fruits. These processes have been traditionally conducted in multistage batteries of tubular type fermentors.

Bilford and co-workers[8] described a one fermentor process for the continuous production of alcohol from molasses at ambient temperatures. The sugar concentration of the molasses had to be diluted to 12% and some molasses required nitrogen and phosphate supplementation. The majority of the sugar was fermented within 7 hr using this system. The fermentation was not examined over a long duration and alcohol yields were not reported. However, the authors did state that yields were comparable with 50 hr batch fermentations. One of the problems of continuous molasses fermentation is that the alcohol yields are generally 1 to 4% lower than the yeast reuse method. Dyr and Krumphanzl[19] described the design and testing of a continuous vertical fermentor using molasses mash and pitching to achieve better yields. This system produced an average yield of 58 and 59% alcohol from the intake sugar with a residence time slightly longer than 24 hr. The authors recommend the yeast, *Saccharomyces cerevisiae,* be regenerated by a short aeration period.

Yarovenko[77] reported thoroughly the investigation of plant scale continuous alcohol fermentation by yeast and butanol-acetone fermentation by bacteria. These processes adopted in plants were conducted on raw material starch and utilized a battery of fermentors efficiently. Optimum conditions required a mass exchange in the main fermentors of 1.3/day for the yeast and 2.5/day for the butanol-acetone bacteria. Alcohol productivity was reported to be increased by 15% compared to the batch process. Carbohydrate utilization however, increased by only 0.45% demonstrating an increased efficiency in end product formation. Similar results were reported for continuous butanol-acetone production. The carbohydrate utilization increased 2.4% which resulted in a 20% productivity increase compared to the semicontinuous battery system. One problem that hindered continuous alcohol production from distillery mashes involved the pretreatment for grain conversion. Grain pretreatment required the use of large pressure cookers where the grain was exposed to temperatures between 100 to 120°F at pH 5.5, and then a temperature over 300°F for a short period of time. Following cooling, barley malt was added and conversion was carried out at 145°F for 30 to 60 min. The converted mash was then pumped through coolers and added to the fermentors. This standard pretreatment used for grain conversion was adequate for batch fermentation, but needed to be modified for feasible use as a continuous process. Gallagher et al.[24] and Unger et al.[73] reported the development of fast grain conversion processes that could be used for continuous alcohol fermentation. Corn cooks could be prepared normally and cooled to a conversion temperature of 62.8°C. Following additions of barley malt, various conversion ties were tested. Compared to normal times, 1 and 5 min. conversion times produced smaller amounts of converted sugar, but the alcohol yield was somewhat higher.[24] The system described by Unger et al.[73] utilized flash conversion for the saccharification of the mash and achieved comparable results. The use of acid-hydrolyzed mash insured complete starch hydrolysis and also provided a sterile mash. Alcohol yields from both batch and continuous fermentations of acid-treated corn mash were found to be depressed compared to malt converted mash fermentations.[61] A small scale continuous alcohol production unit was developed by Altsheler and co-workers[4] for the production of fuel alcohol. The unit was designed for rapid cooking and acid hydrolysis. The fermentation cycle was 11 hr. demonstrating a production rate of 5 gal/day.

Investigations into the continuous production of lactric acid date back to the 1930s.[58,59,76] The nonsterile fermentation of raw whey with continuous feeding of calcium carbonate by lactobacilli at a pH held between 5.0 and 5.8, and a temperature of 43°C was successful at a 24 hr residence time. After the fermented whey was collected the temperature was increased to 100°C to precipitate the milk proteins. Addition of lime to the solution converted the lactic acid to calcium lactate. The coagulum was removed by filtration, and crystallization of the liquid yielded calcium lactate. Lactic acid could be produced directly by the addition of sulfuric acid to the coagulated whey, and evaporation of the filtered solution can be accomplished to achieve the desired lactic acid concentration. This process produced 80% of the theoretical lactic acid yield.

The kinetics of lactic acid production was studied by Kempe et al.[37] with *Lactobacillus delbrueckii*. The organism grows well at pH 5.5 producing 97% of the total acid as lactic acid. The fermentation rate at controlled pH and temperature was constant and predictable under continuous nutrient additions. Low nutrient additions reduce side reactions and increase lactic acid yields from carbohydrates, however, low nutrient concentrations also reduce the rate of fermentation. The use of low nutrient concentration added continuously achieves a high overall rate of lactic acid production. Mil'ko et al.[44] also examined continuous lactic acid production by *L. delbrueckii* at different flow rates. At a rapid growth rate only 57% lactic acid end product was produced resulting in a production of 6g/ℓ/day. The yield could be increased to 90% at slower flow rates, however, only 1.5g/ℓ/day was achieved.

Recently, Griffith and Compere[26] suggested the use of a continuous fixed film system as a pulping industry waste water treatment process to reduce sugar content in the effluent and produce lactic acid as a commercial end product. The fixed film unit was seeded with lactobacilli and lactose fermenting yeasts and operated using wood molasses substrates. The amount of acid produced from this fermentation was approximately 3%.

A variety of reports have appeared dealing with the biological production of growth-associated organic acids. *Propionibacterium acidipropionici (arabinosum)* is a bacterium that grows slowly, but produces propionic acid as an end product. The organism has been cultivated successfully on waste liquor from sulfite wood-pulp production on the bench level. The conversion of fermentable sugars to volatile acids was 83 to 86.5% efficient at a 55 hr residence time using recycled liquid. Both propionic and acetic acid were produced at a ratio of 2:1.[42] *Micrococcus glutamicus* (formerly *Corynebacterium glutamicum)* has been used in a continuous two-stage fermentation for the production of L-glutamic acid. The yield of L-glutamic acid was 51 g/ℓ.[53] Manganese and biotin may be very important in the initiation and production of glutamic acid.[72] Fumaric acid has been produced at significant levels in batch fermentation by *Rhizopus* sp. at the bench scale level. Substrates successfully employed have been glucose, sucrose, and partially inverted sucrose from molasses. The fermentation time is slow, ranging from 3 to 8 days, and dependent upon the sugar concentration, but 75 to 80% of the total acids produced was fumaric. Optimal yields required appropriate levels of nitrogen, zinc, magnesium, phosphorus, iron, corn steep liquor, and methanol.[57] Innovative fermentative concepts have been used-two-stage fermentations for the production of fumaric acid followed by biological conversion to aspartic acid[30] and succinic acid.[31,68] The initial product, fumaric acid, was produced by *Rhizopus* during the first stage of the fermentation. The second stage of the fermentation was accomplished by the addition of the appropriate bacterium. Very good conversion yields were achieved within 24 and 48 hr for the production of asparic acid and succinic acid, respectively. Such novel ideas may play an important role in the future for large scale production of growth associated products in mixed cultures or two-stage continuous processes.

B. Production of Non-Growth Associated Products

The biological production of antibiotics is relatively young in developmental history, however, it is of significant importance to the fermentation industry. Three major classes of industrially important antibiotics, aminoglycosides,[45] B-Lactamas,[25] and macrolides[41] have been recently reviewed. The continuous production of antibiotics has been successfully conducted at the laboratory and pilot plant scale, however, to date continuous cultivation has not been introduced for commercial production.[29,47] Bungay[11] critically examined the possibility of converting the batch production of penicillin G to a continuous fermentation using a 32,000 ℓ system. The cost of the culture medium and the penicillin yield were significant factors that contributed to the operating cost of such a system. The author recommended the use of a multi-stage system as an appropriate means of achieving the highest penicillin yields.

The production of penicillin is probably one of the most important and best understood antibiotic fermentations. Tornqvist and Peterson[71] using high yielding strains of *Penicillium chrysolgenum,* analyzed selected factors that affected penicillin yield. The strains were examined in shake flask experiments using a medium of corn steep solids and lactose. In summary, penicillin yields as high as 1150 units/ℓ were achieved with this medium. Small aliquots of oil added to the culture at 24 hr intervals were found to stimulate product formation by acting as an additional nutrient source. Optimal results were achieved in a medium consisting of 3% corn steep solids and 5% lactose. The authors noted penicillin production was accompanied by a decrease from 9 to 6% of the mycelial nitrogen content. They concluded a mycelium of high nitrogen content was conducive to higher yields of penicillin.

Pirt and Callow[50] placed the organism into continuous culture to specifically examine growth in relation to penicillin production. The fermentation was conducted for a period of 1000 hr. The authors noted that penicillin production was proportional to the mycelial concentration, the residence time, and the metabolic quotient (units of penicillin produced per hour per gram dry weight). The penicillin metabolic quotient remained constant in continuous culture when the dilution rate and dry weight values were varied. This is in contrast to batch culture studies where the metabolic quotient falls near zero within 100 to 150 hr. Batch cultures did, however, produce a higher maximum penicillin concentration over continuous cultivation. Additional investigations by Pirt and Righelato[51] supported the suggestion that penicillin production was dissociated from the growth of the organism. In addition, the production of penicillin was extremely sensitive to carbon dioxide concentration. This ruled out the use of calcium carbonate as an inexpensive means to control pH in the product formation period.[52] Since product formation is dissociated from mycelial growth, the use of the two-stage or multistage fermentation design of the type described by Reusser[54] would be the most desirable. As long as a suitable substrate is added at a low concentration to support the maintenance requirements of the mycelia and prevent deterioration, the product yield would be maximized.

The use of inexpensive substrates for use in commercial penicillin fermentation has been extensively investigated since the early nutritional work. A wide variety of carbohydrates were examined, however, lactose gave superior results due probably to its slow rate of utilization. Commercial producers find the cost of the conventional purified carbohydrates to be excessive and have resorted to less expensive crude substrates (such as hydrol molasses) for commercial penicillin production. In a recent effort to find a suitable substrate, Newman et al.[46] compared Cultrate 6300 Fermentation Substrate to standard glucose-lactose and hydrol molasses media. Over similar fermentation times the penicillin yield from the Cultrate 6300 medium was found to be 83% higher than hydrol molasses and 11% over the glucose-lactose standard medium. It would be interesting to note if similar yields could be obtained using a continuous fermentation.

A large number of other commercially important antibiotics have been found to be produced following the active growth cycle of the population. Of these, only a few have been examined in continuous culture. One such antibiotic is novobiocin, which is produced by *Streptomyces niveus*. Smith[65] reported the production of 550 mg of novobiocin per liter following batch growth on a medium of 4% sucrose and 4% distillers solubles. Reusser[55] examined a nine-stage continuous fermentation system for the production of novobiocin. The organism had been shown to lose novobiocin producing ability within 10 to 25 days when grown in batch culture. Therefore, two interchangeable stages were employed at the front end of the continuous system and by alternating from one to the other at weekly intervals, newly cultured cells were continuously added to the remaining stages. Most runs were conducted for 2 to 3 months, however, maximum output was maintained for 33 days at a level of 350 to 700 mg novobiocin per liter present in the last stage.

Two-stage continuous fermentation has been successful for the production of chlortetracycline by *S. aureofaciens*.[63] The antibiotic concentration was found to increase up to 50 hr of cultivation and then diminish considerably. The chlortetracycline concentration in the first and second stages prior to the decrease was 1250 and 2500 mg/ℓ, respectively. However, when the phosphorus concentration of the medium was lowered to 30 mg/ℓ no decrease in antibiotic concentration was observed during the entire course of the fermentation (200 hr). In addition, results from batch experiments conducted with the gramicidin-S producing bacterium, *Bacillus brevis*, suggested a two-stage continuous fermentation may prove successful.[74]

Recently, results were presented on erythromycin production by *S. erythreus* grown semicontinuously on a complex medium. The organism was grown in the first stage fermentor for 48 hr and then part of the first stage fermentor broth was added to the second stage. Sterile medium was then added continuously to the first stage. This cycle was repeated daily, resulting in a 79% increase in erythromycin productivity over the batch process. In addition, the authors pointed out that conventional batch equipment could be easily adapted for the semicontinuous process in scale up operations.[10]

Commercial production of enzymes is a relatively young field also; however, there are a variety of important representatives. Aunstrup[5] has reviewed the major industrially important enzymes. A number of improvements have been developed for the production of enzymes, but no industrial or pilot plant continuous operations have been reported in the literature. A few reports of continuous production of enzymes at the laboratory level have appeared, which suggest that the continuous production of enzymes is feasible.

Proteases produced by bacilli at the commercial level have become extremely important as detergent additives and in the brewing industry. Many of these enzymes are produced in substantial amounts during the stationary phase of batch grown cells. The regulatory mechanism of protease production is unknown and rather complex. Data provided from batch experiments however, led Jensen[35] to examine a two-stage continuous system for the production of protease by *Bacillus subtilis.* The first stage was designed to function as a continuous inoculum for the second stage. Protease production occurred in single-stage chemostats at low dilution rates, but became negligible at more rapid dilution rates. In the two-stage system, the second stage was three times the volume of the first stage, which increased the residence time in the final stage. This design permitted a 24 hr fermentation time and a substantial flow rate. Protease activity assayed from the second-stage effluent was equal to 60% of the activity present in 48 hr batch cultures. The system was operated without interruption successfully over a 26 day period.

Pullulanase is an industrially important debranching enzyme that hydrolyzes α-1,6 glycosidic linkages when α-1,4 linkages are present on adjacent sides. Chemostat cultures of *Klebsiella aerogenes* produced an optimum amount of enzyme when an inducer served as a carbon limiting nutrient. The inducers examined were maltose, maltotriose, pullulan, and acid hydrolysates of wheat starch. Pullulanase remained cell bound in all cases.[27] Increasing the culture biomass did not affect the optimum level of enzyme produced; however, smaller amounts of enzyme remained bound firmly to the cell. Nearly 30% of the total enzyme activity was cell-free. It was interesting that a sudden change from aerobic to anaerobic conditions caused a rapid release of bound enzyme. The rapid release of enzyme by conversion to anaerobiosis was not a steady-state phenomenon. The same values established during the aerobic conditions became reestablished under anaerobic conditions. Also, pullulanase was not induced when *K. aerongenes* was cultured under anaerobic conditions from the start.[28]

A number of other amylolytic enzymes of industrial importance have been described.[5] All of these are currently produced by batch processes, and attempts at production by single stage continuous methods have resulted in marginal success. Fencl et al.[22] examined α-amylase production by *B. subtilis* and offered an explanation as to why many metabolites cannot be successfully produced continuously. In a continuous process they observed that the enzyme concentration increased to 40 units per milliliter and then decreased suddenly. A short oscillation period followed and the concentration stabilized at a value 70 to 80% lower than batch culture values. The authors offered the explanation that different environmental conditions are required for the transcription of specific messenger RNAs and for the synthesis of a corresponding enzyme. Additional investigations demonstrated that the maximum synthesis of the RNA occurred during the exponential growth phase; however, the maximum accumulation of product occured near the stationary phase.

Contrary to the production of α-amylase by *B. subtilis,* Davis et al.[15] reported an optimistic single-stage continuous production of the enzyme by *B. stearothermophilus.*

The enzyme is produced during the exponential stage of the organism in batch culture and is extremely heat stable. Comparison of the amylase yields from batch and continuous culture revealed three times as much enzyme was produced in a given time in continuous culture with only a small increase in medium consumption.

Imanaka et al.[33] reported the optimization of α-galactosidase production by a *Monascus* sp. The enzyme is intracellular and induced by galactose, melibiose, raffinose, or stachyose. Glucose was used as an inexpensive growth substrate followed by enzyme induction by the addition of galactose. Glucose concentrations greater than 2.25×10^{-4} gram/mℓ were found to repress galactose utilization and α-galactosidase production. Further optimization was achieved by conversion from a single stage to multistage continuous process. The most productive design examined was a combination of two stirred tanks followed by a tubular fermentor. Glucose was fed continuously as the growth substrate in the first fermentor and the temperature was controlled at 30°C. Galactose was added as the inducer to the third tank and the temperature was increased to 35°C. This fermentation system increased the α-galactosidase yield 25.6% over the single-stage continuous process.[34]

Commercial fermentation can be economically more feasible if more than one end product can be produced from a single fermentation. Sikyta and Fencl[64] described the simultaneous induction of β-galactosidase and tryptophanase in *Escherichia coli.* Simultaneous induction of these enzymes during batch cultivation resulted in depressed values of tryptophanase in the inducible strain. The specific activity of β-galactosidase in the presence of its inducer then decreased to a basal level. On the other hand, in continuous culture, the induced enzymes maintained a constant level that exceeded batch culture values.

Glucose isomerase and invertase are two enzymes of commercial interest and have been studied in continuous culture. Glucose isomerase is used by manufacturers to produce sugar-like products from starch hydrolysates. Diers[16] examined glucose isomerase production by an atypical thermophilic variant of *B. coagulans.* Growth limitation in continuous culture by inorganic compounds with glucose in excess resulted in repression of enzyme synthesis. Carbon limitations and dual carbon oxygen limitation were advantageous to enzyme productivity. Invertase production in continuous culture is unique in that there is an optimum growth rate resulting in maximum enzyme production in *Saccharomyces carlsbergensis.*[69,70]

C. Production of Transformation Products

Fermentations resulting in the bioconversion of chemicals are of major industrial importance. These fermentations are similar to many antibiotic fermentations in that they are multistage processes carried out in batch at the industrial level. The first stage functions as a growth process for the generation of cells. Once the population has been established, the environmental conditions are altered — leading towards product formation in the second stage. A variety of these products have been examined at the bench scale level in continuous systems; however, industrial development will probably require an in depth understanding of the physiological processes involved in product formation.

The microbial transformation of chemicals has been of great interest in the last decade due to the convenience and specificity of these processes. Both whole cells and enzymes have been examined successfully and current reviews are available that discuss the bioconversion of nonsteroid organic molecules,[38,62] pharmaceuticals[7] and steroids[12], and pesticides.[9,36] Even though bioconversion fermentation has generated tremendous interest, very few papers can be found in the literature examining continuous processes. However, the few reports of continuous steroid transformation examined in the early 1960s may provide a foundation for future interest in continuous bioconversion.

The work of Dulaney and Stapley[18] is typical of the observations first noted concerning biotransformation of steroid compounds. A strain of *Curvularia lunata* was isolated from the soil, which could effect the low yield conversion of 11-deoxy-17-hydroxycorticosterone to hydrocortisone after 24 to 48 hr of growth in complex media batch culture. The yields could be increased significantly by the use of synthetic media; however, nutritional differences were noted between the transformation of steroids by *Curvularia* and a previous study using *Aspergillus ochraceus.*[17] It was interesting to note that increasing additions of 11-deoxy-17-hydroxycorticosterone resulted in greater amounts of hydrocortisone converted, but the yields decreased. The best yields achieved in synthetic media were 50%.

Continuous microbiological transformation of steroids was first examined by Mateles and Fuld[43] and Reusser et al.[56] Mateles and Fuld[43] utilized the typical two-stage continuous fermentation where growth of *A. ochraceus* was achieved in the first stage, and progesterone was continuously added to the second stage for conversion. The two-stage design was optimal for this conversion since progesterone was found to inhibit cell growth. Conversion in the second resulted in 36 mg of 11-hydroxyprogesterone per gram of mycelium per 8 hr contact time. Reusser et al.[56] conducted the same transformation using *Rhizopus nigricans,* and also the bioconversion of pregnadiene to pregnatriene using *Septomyxa affinis* in an eight-stage fermentation process. Following growth in the first stage, the steroids for transformation were fed continuously into the second stage. Both conversions were achieved within a 5 hr period resulting in yields of 50 to 60%. Though an eight-stage system was employed, it was suggested that a two-stage system would be optimal, and examination of all operational variables should make this a very feasible process of fermentation.

A large number of continuous fermentation processes have been described as two-stage or multistage systems, where an actively growing population is established in the first stage and the environmental conditions are adjusted in the remaining stages resulting in product formation. The amount of time required to establish the population and, more importantly, the steady-state population density are two factors that influence the yield and subsequently the economic feasibility of the fermentation. Srinivasan et al.[66] described a technique for extending the period of exponential growth to achieve a technique for extending the period of exponential growth to achieve a high cell density culture of a *Cellulomonas* sp. The organism appears to be extremely sensitive to environmental influence by concentrations of nutrients including trace elements. They had noted previously that specific concentrations of certain trace elements were required to maintain an optimum growth rate, however, slight increases in concentration resulted in a decline in growth rate. After the construction of a defined medium, they were able to show that by increasing the concentration of all nutrients at a rate near the maximum growth rate of the organism, very high yields could be produced in a short period of time. Optimum growth conditions are maintained for a longer period of time because the amount of each given nutrient per organism remained constant during the entire growth phase. By this method biomass yields were increased from 2.1 to near 30 g/ℓ dry weight within a 10 hr growth phase. Significant increases in yield by this method were also reported for four different bacilli.[67] The future deveopment and industrial application of this kind of cultivation technique may prove to significantly increase product yields of growth related products, secondary metabolites, and bioconversion products.

III. CONTINUOUS BEER PRODUCTION

The brewing industry has maintained interest in continuous fermentation for well over 2 decades. Many of the technological problems associated with continuous cultivation have been overcome by advances in beer fermentation. Nevertheless, the conversion from batch processes to continuous and semicontinuous processes has experienced a number of disappointments. The most significant difficulty may be the inability to maintain a specific product quality because of the variability present in the raw materials. Thus, only a small percentage of the beer is produced continuously on a worldwide level.

Hough et al.[32] have described the traditional steps for the production of beer. The principal substrate is the brewers wort, which is formed by crushing malted barley and adding water under conditions that allow enzymatic conversion of the malt. The mixture is filtered and the extract is boiled with hops, cooled, and is inoculated with the yeast. Two types of yeasts are used in common beer fermentations. The ale yeasts rise to the surface near the end of the fermentation, whereas the lager yeasts sediment. The beer produced is then passed through a variety of postfermentation treatments to achieve a desirable product of consistent quality.

Two types of continuous beer producing processes have been reviewed.[29,32] The first of these deals with stirred tank processes. The specifics of this fermentation vary; however, usually two stirred tanks are joined in series followed by an unstirred conical tank for yeast separation and product clarification. The first stirred tank functions for the production of actively growing yeast. The vessel is fed with aerated wort, yeast, and recycled fermenting wort. The fermentation occurs in the second stirred vessel. Hough et al.[32] also described the three vessel cascade system of the Wayney Breweries. Briefly, the first two vessels are fermenting vessels and the third is the separator. The wort is held at 0°C and oxygenated to 10 ppm dissolved oxygen prior to entry into the first vessel. The specific gravity of the wort is reduced in the first vessel from 1.040 to 1.010. The second vessel maintains a higher yeast concentration and the specific activity is further reduced from 1.019 to 1.011. The residence time for the system, including separation, is 15 to 20 hr using worts in the range of 1.035 to 1.040. Usually the system can be run for 3 month periods between cleaning shutdowns.

The second type is the most successful continuous system functioning to date. The continuous tower fermentor described by Klopper et al.[39] and Royston[60] is a closed system that can be run for long periods of time between cleanings. Immediate interest was established throughout the world, as indicated by registered patents.[2,3,21,78,79] Hough et al.[32] have described the use of the APV system that is summarized here. The tower is carefully cleaned and steam sterilized. Then the tower is partially filled with aerated wort and the yeast inoculum. Aeration occurs continuously, and at intervals over a 9 day period wort is added. The flow rate of the vessel is gradually increased to an average residence time of 4 to 5 hr by the introduction of wort at the base of the tower. The gradual increase occurs over an 18 to 21 day interval to establish a very heavy yeast concentration near the bottom of the tower where a plug is formed. The wort flows upwards through the plug where a rapid fermentation begins. The plug remains stable, and a perforated plate near the base prevents channeling of the wort as it passes upwards and is progressively fermented, resulting in a drop in pH and specific gravity. The fermenting liquid in the upper region flocculates and the yeast clump. The specific gravity of the yeast is greater than the wort and the yeast is separated by gravity. Yeast strains are selected to give adequate separation. Once in operation it may take up to 3 months to achieve the product consistency desired, but the steady-state operation can continue 12 months or longer.

IV. FAR RANGE DEVELOPMENT

Throughout the years, scores of investigations have been conducted that have aided in the development of sound batch fermentations of industrial importance. The full potential for most batch fermentations has been achieved using the inexpensive crude substrates and nutrients desired by industry. Even though the theoretical physical and economic advantages of continuous processes are known, they are seldom achieved following conversion from batch fermentations to fully or semicontinuous operations. Often this is due to poor understanding of the physiology of the organisms at steady-state conditions or their sensitivities to environmental factors. Continued research using continuous culture as a tool will need to be conducted on all levels of research before the full level of potential can be reached in the commercial process.

The yield of any fermentation product is dependent upon the concentration of all nutrients and the microbial population in the medium. Usually at the end of the fermentation the cells need to be separated from the product as a purification step. Recently, a significant thrust has been made into some novel aspects of continuous product formation that may prove to be of economic importance in the future. The immobilization of cells and enzymes enables concentration to increase fermentation efficiency and reduce reactor size. Once immobilized a suitable nutrient supply or reaction mixture may be passed through the matrix continuously. These aspects are currently receiving widespread attention, and it is impossible to discuss these developments in depth in this paper. However, recent reviews are available that cover these topics thoroughly.[1,6]

In reference to cells, the term immobilization refers not only to a process where cells are entrapped in a matrix, but also any means whereby cells are recovered and reused. Cell pellets and flocculants can be used continuously in column reactors. Abbott[1] mentions that some properties of immobilized cells may vary from cells free in suspension, and specific advantages may be associated with specific fermentations. Nutrient limitation or toxic materials that effect a washout in conventional continuous process may not severely hinder fermentations conducted with cells immobilized in a matrix. However, films, flocculated, or pelleted cells may still be subject to these stresses in continuous operations resulting in a loss in biomass. It must also be kept in mind that catabolic enzymes are also present in immobilized cells and some quantity of the product is lost through degradation. This is not the case of immobilized enzymes, where the specific enzymes desired is covalently coupled to an inert support matrix.

Immobilized cells have been effectively used in activated sludge and trickling filters for the reduction of Biochemical Oxygen Demand (BOD) and Chemical Oxygen Demand (COD) in sewage treatment processes. Recent reports suggest that similar processes may be useful in the treatment of industrial wastewaters. Etzel and Kirsch[20] reported the use of a fiber wall reactor inoculated with pentachlorophenol oxidizing bacterial cells for the reduction of pentachlorophenol from industrial effluents. A synthetic wastewater containing various concentrations of the pollutant and an authentic wood-preservative wastewater was examined by continuous addition to the reactor. The system proved effective in removing as much as 80% of the biocide and the standard treatment parameters were met at a 97% efficiency. Compere and Griffith[13] described the use of a continuous fixed-film bioreactor for the successful removal of nitrate from high-strength industrial nitrate waste. The choice of a suitable carbon source appeared to be the most important consideration for the effective removal of nitrate. Although the carbon source plays a significant role, nitrate reduction could be achieved at values greater than 90%.

V. SUMMARY

The purpose of this chapter was to discuss the industrial applications of continuous fermentation processes. As pointed out, there are only a few examples of successful operations being conducted currently. However, continued investigation and development of existing biological processes may increase the role played by continuous cultivation in the biological production of materials. Presently, continuous cultivation shows vast potential, which will need to become a reality to deal with the increased population pressure and the decreased availability of raw materials.

REFERENCES

1. **Abbott, B. J.**, Immobilized cells, in *Annual Reports on Fermentation Processes,* Vol. 1, Perlman, D., Ed., Academic Press, New York, 1977, 205.
2. Allied Breweries Ltd. Manufacture of beer by continuous fermentation. Fr. 1,422-477 (Cl. C12b,c). Appl. Jan. 5, 1965a; 11pp.
3. Allied Breweries Ltd. Continuous fermentation. Neth. 6,500,043 (Cl. C12c). Appl. Jan. 5, 1966; 21 pp.
4. **Altsheler, W. B., Mollet, H. W., Brown, E. H. C., Stark, W. H., and Smith, L. A.**, Design of a two-bushel per day continuous alcohol unit, *Chem. Eng. Prog.*, 43, 467, 1947.
5. **Aunstrup, K.**, Enzymes of industrial interest: traditional products, *Annual Reports on Fermentation Processes,* Vol. I, Perlman, D., Ed, Academic Press, New York, 1977, 181.
6. **Bernath, F. R., Venkatasubramanian, K., and Vieth, W. R.**, Immobilized enzymes, in *Annual Reports on Fermentation Processes,* Vol. 1, Perlman, D., Ed., Academic Press, New York, 1977, 235.
7. **Beukers, R., Marx, A. F., and Zuidweg, M. H. J.**, "Microbial conversion as a tool in the preparation of drugs," *Drug Design,* Vol. III, Ariens, E. J., Ed., Academic Press, New York, 1973, 1.
8. **Bilford, H. R., Scalf, R. E., Stark, W. H., and Kolachov, P. J.**, Alcoholic fermentation of molasses. Rapid continuous fermentation process, *Ind. Eng. Chem.*, 34, 1406, 1942.
9. **Bollag, J. M.**, Microbial transformation of pesticides, *Adv. Appl. Microbiol.*, 18, 75, 1974.
10. **Bosnjak, M., Holjevac, M., and Johanides, V.**, Erythromycin Biosynthesis in Semicontinuous Culture of *Streptomyces erythreus,* 6th Int. Symp. on Continuous Culture of Microorganisms, St. Catherine's College, Oxford, July 1975, 20.
11. **Bungay, H. R.**, Economic definition of continuous fermentation goals, *Biotechnol. Bioeng.* 5, 1, 1963.
12. **Charney, W. and Herzog, H. H.**, *Microbial Transformation of Steroids,* Academic Press, New York, 1967.
13. **Compere, A. L. and Griffith, W. L.**, Continuous fixed-film denitrification of high-strength industrial nitrate wastes, *Dev. Ind. Microbiol.*, 18, 717, 1977.
14. **Dawson, P. S. S.**, Continuous fermentations, *Annual Reports on Fermentation Processes,* Vol. I, Perlman, D., Ed., Academic Press, New York, 1977, 73.
15. **Davis, P. E., Whitaker, A., and Cohen, D. L.**, The Production of Amylase by *Bacillus stearothermophilus* in Continuous Culture, 6th Int. Symp. on Continuous Culture of Microorganisms, St. Catherine's College, Oxford, July 1975, 20.
16. **Diers, J.**, Glucose isomerase in *Bacillus coagulans, Continuous Culture 6,* Dean, A. C. R., Ellwood, D. C., Evans, C. G. T., and Melling, J., Eds., John Wiley & Sons, New York, 1976, 208.
17. **Dulaney, E. L., Stapley, E. O., Hlavac, C.**, Hydroxylation of steroids, principally progesterone, by a strain of *Aspergillus ochraceus, Mycologia,* 47, 464, 1955.
18. **Dulaney, E. L. and Stapley, E. O.**, Studies on the transformation of 11-deoxy-17-hydroxycorticosterone to hydrocortisone with a strain of *Curvularia lunata, Appl. Microbiol.*, 7, 276, 1959.
19. **Dyr, J. and Krumphanzl, V.**, Continuous alcohol fermentation in a new apparatus of the vertical type, in *Continuous Cultivation of Microorganisms,* Malek, I., Beran, K., and Hospodka, J., Eds., Academic Press, New York, 1964, 199.
20. **Etzel, J. E. and Kirsch, E. J.**, Biological treatment of contrived and industrial wastewater containing pentachlorophenol, *Dev. Ind. Microbiol.*, 16, 287, 1975.
21. Falstaff Brewing Corp. A continuous fermentation for production of beer. Neth. 6,604,530 (Cl. C12c). Appl. Oct. 6, 1966; U.S. Appl. April 5, 1965; 14 pp.
22. **Fencl, Z., Ricica, J.,and Kodesova, J.**, The use of the multistage chemostat for microbial product formation, *J. Appl. Chem. Biotechnol.* 22, 405, 1972.
23. **Freeman, G. G. and Morrison, R. I.**, Production of 2,3-butylene glycol by fermentation of molasses, *J. Soc. Chem. Ind. London,* 66, 216, 1947.
24. **Gallagher, F. H., Bilford, H. R., Stark, W. H., and Kolachov, P. J.**, Fast conversion of distillery mash: for use in a continuous process, *Ind. Eng. Chem.*, 34, 1395, 1942.

25. **Gorman, M. and Huber, F.**, β-lactam antibiotics, in *Annual Reports on Fermentation Processes,* Vol. I, Perlman, D., Ed., Academic Press, New York, 1977, 327.
26. **Griffith, W. L. and Compere, A. L.**, Continuous lactic acid production using a fixed-film system, *Dev. Ind. Microbiol.,* 18, 723, 1977.
27. **Hope, G. E. and Dean, A. C. R.**, Pullulanase synthesis in *Klebsiella (Aerobacter) aerogenes* strains growing in continuous culture, *Biochem. J.,* 144, 403, 1974.
28. **Hope, G. E. and Dean, A. C. R.**, Pullulanase synthesis in *Klebsiella aerogenes* growing at high biomass levels in maltose-limited chemostat culture, *J. Appl. Chem. Biotechnol.,* 25, 549, 1975.
29. **Hospodka, J.**, Industrial application of continuous fermentation, *Theoretical and Methodological Basis of Continuous Culture of Microorganisms,* Malek, I., and Fencl, Z., Eds., Academic Press, New York, 1966, 495.
30. **Hotta, K. and Takao, S.**, Conversion of fumaric acid fementation to aspartic acid fermentation by the association of Rhizopus and bacteria. II. Production of aspartic acid by the combination of *Rhizopus* and *Proteus vulgaris, J. Ferment. Technol.,* 51, 12, 1973.
31. **Hotta, K. and Takao, S.**, Conversion of fumaric acid fermentation to succinic acid fermentation by the association of *Rhizopus* and bacteria. IV. Production of succinic acid by the association of *Rhizopus arrhizus* and *Aerobacter aerogenes* and the mechanism of the conversion of the fermentation, *J. Ferment. Technol.,* 51, 26, 1973.
32. **Hough, J. S., Keevil, C. W., Maric, V., Philliskirk, G., and Young, T. W.**, Continuous culture in brewing, in *Continuous Culture 6,* Dean, A. C. R., Ellwood, D. C., Evans, C. G. T., and Melling, J., Eds., John Wiley & Sons, New York, 1976, 226.
33. **Imanaka, T., Kaieda, T., Sato, K., and Taguchi, H.**, Optimization of β-galactosidase production by mold, *J. Ferment. Technol.,* 50, 633, 1972.
34. **Imanaka, T., Kaieda, T., and Taguchi, H.**, Optimization of β-galactosidase production in multi-stage continuous culture of mold, *J. Ferment. Technol.,* 51, 431, 1973.
35. **Jensen, D. E.**, Continuous production of extracellular protease by *Bacillus subtilis* in a two-stage fermentor, *Biotechnol. Bioeng.,* 14, 647, 1972.
36. **Kearny, P. C. and Kaufman, D. D.**, *Degradation of Herbicides,* Marcel Dekker, New York, 1969.
37. **Kempe, L. L., Gillies, R. A., and West, R. E.**, Acid production by homofermentative lactobacilli at controlled pH as a tool for studying the unit process of fermentation, *Appl. Microbiol.* 4, 175, 1956.
38. **Kieslich, K.**, *Microbial Transformations of Nonsteroid Cyclic Compounds.* Georg Thieme, Stuttgart and John Wiley & Sons, New York, 1975.
39. **Klopper, W. J., Roberts, R. H., Royston, M. G., and Ault, R. G.**, Continuous fermentation in a tower fermentor, *Eur. Brew. Conv., Proc. Congr.,* 238, 1965.
40. **Long S. K. and Patrick, R.**, Production of 2,3-butylene glycol from citrus wastes, *Appl. Microbiol.,* 9, 244, 1961.
41. **Majer, J.**, Macrolide antibiotics, *Annual Reports on Fermentation Processes,* Vol. I, Perlman, D., Ed., Academic Press, New York, 1977, 347.
42. **Martin, M. E., Wayman, M., and Graf, G.**, Fermentation of sulfite waste liquor to produce organic acids, *Can. J. Microbiol.,* 7, 341, 1961.
43. **Mateles, R. I. and Fuld, G. J.**, Continuous hydroxylation of progesterone by *Aspergillus ochraceus, Antonie Van Leeuwenhoek J. Microbiol. Serol.,* 27, 33, 1961.
44. **Mil'ko, E. S., Sperelup, O. V., and Rabotnova, I. L.**, Lactic acid production by *Lactobacillus delbrueckii* in continuous fermentation, *Z. Allg. Mikrobiol.,* 6, 297, 1966.
45. **Nara, T.**, Aminoglycoside antibiotics, in *Annual Reports on Fermentation Processes,* Vol. I, Perlman, D., Ed., Academic Press, New York, 1977, 299.
46. **Newman, H., Skole, R. D., Hogu, J., and Rizzoto, A. B.**, The development of a new inexpensive carbon source for antibiotic production, *Dev. Ind. Microbiol.,* 16, 375, 1975.
47. **Nyiri, L. K. and Charles, M.**, Economic status of fermentation process, in *Annual Reports on Fermentation Processes,* Vol. I, Perlman, D., Ed., Academic Press, New York, 1977, 365.
48. **Perlman, D.**, Production of 2,3-butylene glycol from wood hydrolysate, *Ind. Eng. Chem.,* 36, 803, 1944.
49. **Pirt, S. J. and Callow, D. S.**, Exocellular product formation by microorganisms in continuous culture. I. Production of 2,3-butanediol by *Aerobacter aerogenes* in a single stage process, *J. Appl. Bacteriol.,* 21, 188, 1958.
50. **Pirt, S. J. and Callow, D. S.**, Studies of the growth of *Penicillium chrysogenum* in continuous flow culture with reference to penicillin production, *J. Appl. Bact.,* 23, 87, 1960.
51. **Pirt, S. J. and Righelato, R. C.**, Effect of growth rate on the synthesis of penicillin by *Penicillium chrysogenum* in batch and chemostat cultures, *Appl. Microbiol.,* 15, 1284, 1967.
52. **Pirt, S. J. and Mancini, B.**, Inhibition of penicillin production by carbon dioxide, *J. Appl. Chem. Biotechnol.,* 25, 781, 1975.

53. **Reisman, H. B., Gore, J. H., and Garden, C. H.,** L-Glutamic acid by continuous fermentation, Merck and Co., Inc. U.S. Patent Appl. Jan. 5 and Nov. 29, 1965; 11 pp.
54. **Reusser, F.,** Theoretical design of continuous antibiotic fermentation units, *Appl. Microbiol.*, 9, 361, 1961.
55. **Reusser, F.,** Continuous fermentation of novobiocin, *Appl. Microbiol.*, 9, 366, 1961.
56. **Reusser, F., Hoepsell, H. J., and Savage, G. M.,** Continuous microbiological transformation of steroids, *Appl. Microbiol.*, 9, 346, 1961.
57. **Rhodes, R. A., Moyer, A. J., Smith, M. L., and Kelley, S. E.,** Production of fumaric acid by *Rhizopus arrhizus, Appl. Microbiol.*, 7, 74, 1959.
58. **Rogers, L. A. and Whittier, E. O.,** The growth of bacteria and continuous flow of broth, *J. Bacteriol.*, 20, 127, 1930.
59. **Rogers, L. A. and Whittier, E. O.,** The continuous fermentation of whey, *J. Bacteriol.*, 21, 37, 1931.
60. **Royston, M. G.,** Tower fermentation of beer, *Process Biochem.*, 1, 215, 1966.
61. **Ruf, E. W., Stark, W. H., Smith, L. A. and Allen, E. E.,** Alcoholic fermentation of acid-hydrolyzed grain mashes: continuous process, *Ind. End. Chem.*, 40, 1154, 1948.
62. **Sebek, O. K. and Kieslich, K.,** Microbial transformations of organic chemicals, in *Annual Reports on Fermentation Processes*, Vol. I, Perlman, D., Ed., Academic Press, New York, 1977, 267.
63. **Sikyta, B., Slezak, J., and Herold, M.,** Continuous chlortetracycline fermentation, in *Continuous Cultivation of Microorganisms*, Malek, I., Beran, K., Hospodka, J. Eds., Academic Press, New York, 1964, 173.
64. **Sikyta, B. and Fencl, Z.,** Continuous production of enzymes, in *Continuous Culture 6*, Dean, A. C. R., Ellwood, D. C., Evans, C. G. T., and Melling, J., Eds., John Wiley & Sons, New York, 1976, 158.
65. **Smith, C. G.,** Fermentation studies with *Streptomyces niveus, Appl. Microbiol.*, 4, 232, 1956.
66. **Srinivasan, V. R., Fleenor, M. B., and Summers, R. J.,** Gradient-feed method of growing high cell density cultures of *Cellulomonas* in a bench-scale fermentor, *Biotechnol. Bioeng.*, 19, 153, 1977.
67. **Summers, R. J., Boudreaux, D. P., and Srinivasan, V. R.,** Gradient Feed Method of Growing High Cell Density Cultures, Annu. Meeting Am. Soc. Microbiol., New Orleans, La., May 1977.
68. **Takao, S. and Hotta, K.,** Conversion of fumaric acid fermentation to succinic acid fermentation by the association of *Rhizopus* and bacteria. III. Conversion of fumarate to succinate by *Aerobacter aerogenes, J. Ferment. Technol.*, 51, 19, 1973.
69. **Toda, K.,** Invertase biosynthesis by *Saccharomyces carlsbergensis* in batch and continuous cultures, *Biotechnol. Bioeng.*, 18, 1103, 1976.
70. **Toda, K.,** Dual control of invertase biosynthesis in chemostat culture, *Biotechnol. Bioeng.*, 18, 1117, 1976.
71. **Tornqvist, E. G. M. and Peterson, W. H.,** Penicillin production by high-yielding strains of *Penicillium chrysogenum Appl. Microbiol.*, 4, 277, 1956.
72. **Ueda, K., Takahashi, H., and Oguma, T.,** Continuous glutamic acid fermentation. III. Basic conditions of glutamic acid production by continuous multi-stage fermentation, *Hakko Kogaku Zasshi*, 4, 645, 1963.
73. **Unger, E. D., Willkie, H. F., and Blankmeyer, H. C.,** The development and design of a continuous cooking and mashing system for cereal grains, *Trans. Am. Inst. Chem. Eng.* 40, 421, 1944.
74. **Vandamme, E. J. and Demain, A. L.,** Nutritional requirements of the gramicidin S fermentation, *Dev. Ind. Microbiol.*, 17, 51, 1976.
75. **Ward, G. E., Pettijohn, O. G., and Coghill, R. D.,** Production of 2,3-butanediol form acid-hydrolyzed starch, *Ind. Eng. Chem.*, 37, 1189, 1945.
76. **Whittier, E. O. and Rogers, L. A.,** Continuous fermentation in the production of lactic acid, *Ind. Eng. Chem.*, 23, 552, 1931.
77. **Yarovenko, V. L.,** Principles of the continuous alcohol and butanol-acetone fermentation processes, in *Continuous Cultivation of Microorganisms*, Malek, I. Beran, K., and Hospodka, J., Eds., Academic Press, New York, 1964, 205.
78. **Zhukov, A. M.,** Continuous automatic installation for fermentation of wort, U.S.S.R. 178,338 (Cl. C12g). Appl. Dec. 12, 1964.
79. **Ziemann, A. and G. m. b. Ger.** Continuous fermentation of beer, Ger. 1,205,041 (Cl. C12b). Appl. 30 Dec., 1961; 5 pp.

Chapter 6

SINGLE-CELL PROTEIN PRODUCTION FROM METHANE AND METHANOL IN CONTINUOUS CULTURE*

J. W. Drozd and J. D. Linton

TABLE OF CONTENTS

I. Introduction ... 114

II. Importance of Continuous Culture ... 114

III. Some Parameters Important to SCP Production ... 115
- A. Substrate ... 115
- B. Organism ... 115
  - 1. Type ... 115
  - 2. Enrichment ... 116
    - a. Batch Enrichment ... 116
    - b. Continuous Enrichment ... 117
- C. Mixed Cultures or Pure Cultures for SCP Production ... 118
- D. Growth Rate ... 121
- E. Affinity of the Organism for the Substrate ... 124
- F. Yield ... 127
  - 1. Factors Affecting Yield ... 127
    - a. Biochemistry of Methane and Methanol Utilization ... 129
    - b. Growth Physiology ... 131
- G. Nutrient Requirements—Problems of Supply at High Productivity ... 132
- H. Fermenter Design ... 133

IV. Conclusions ... 134

V. Appendix ... 136

Acknowledgments ... 137

References ... 137

* This chapter was submitted in July 1978.

I. INTRODUCTION

The literature on single-cell protein production is rapidly increasing and in any review it is not possible to cover all the published papers in a critical manner. We shall aim to look at some papers we consider to cover important microbiological concepts, or give data which is of relevance to the development of single-cell protein production. It is well to start with a broad definition of the term "single-cell protein"; this was a term introduced at the 1966 meeting at MIT.[1] It refers to the production of unicellular microorganisms and mostly covers bacterial, fungal, and algal growth. In general, the product is destined as a protein supplement for animal feed. It is not a new area of biotechnology, in the first world war and subsequent years quite large amounts of yeast were produced from molasses. The area of research has however recently been given extra impetus because of:

1. The protein shortfall which occurs in many industrialized countries that results in the import and use of large amounts of fish meal and soya protein for animal feedstuff.
2. The variability in world fish catches, especially that of anchovies, and the shortfall in fish meal supplies to an expanding population—fish meal is high in many of the essential amino acids (especially methionine, cysteine, and lysine) required in nonruminant nutrition and can often be replaced by single-cell protein. It is also unclear as to whether the production of soybeans in the U. S. and Brazil can be increased to cope with the demands of an increased world population. Soya protein is lower in the essential amino acids and for a balanced feed must be supplemented with a "high quality" protein such as fish meal or single-cell protein. It is a mistake to think of single-cell protein solely as a replacement for soya protein; it should be regarded as a high quality protein supplement which may partly or totally replace fish meal.
3. The isolation of microorganisms which are capable of growth on naturally occuring hydrocarbons (e.g., bacterial growth on the methane fraction in natural gas), and their derivatives (e.g., bacterial or yeast growth on methanol) — in addition to the isolation of the relevant microorganisms an important development was the demonstration that they could be grown to a high density (in excess of 10 g dry weight ℓ^{-1}) at specific growth rates in excess of 0.1 hr^{-1} in continuous culture.

The above mentioned substrates are all nonrenewable, i.e., they are derived from fossil fuels and there is thus a finite (albeit enormous) supply of them on this planet. There is also a great deal of interest in the use of renewable sources of carbon and energy for SCP production, e.g., the growth of photosynthetic blue-green algae with sunlight and carbon dioxide in areas of high light intensity or the utilization of carbohydrate produced from crops,[2] e.g., starch, as a substrate for SCP production. There is also an interest in the utilization of some industrial wastes, e.g., sulfite liquor from the paper industry for SCP production. Although all of these substrates are of interest we shall deal with the use of natural gas, and methanol as substrates for SCP production although n-paraffins and sulfite waste liquor are also the basis of highly researched SCP processes which have been in development, or which are planned.

II. IMPORTANCE OF CONTINUOUS CULTURE

Apart from waste-water treatment, the large scale production of single-cell protein represents the first major use of continuous culture techniques on an industrial scale. Indeed, several recent advances in the theory and practice of continuous culture have developed because of the recent intensive research on single-cell protein production. A continuous culture offers many advantages over batch culture. It is capable of continuously producing product over a long period of time whereas a batch process would involve a large extra cost because of the amount of time product is not being made. Once set up, a continuous process will produce biomass for a period determined solely by the operators, moreover parameters such as growth rate, substrate concentration and product composition can be carefully controlled. In a batch process there is no control over these parameters and there is only a very short period of time for which the culture is at the required cell density, after which the whole cycle of growth must be repeated. The use of a plant dedicated to single-cell protein production can with the use of a continuous culture process be operated for most of the time available, unlike any similar batch process.

The use of continuous culture makes process control much easier, and in a steady state it enables measurements to be made with great accuracy over a long period of time. This enables process control strategies to be developed and it also enables factors such as yield to be measured with great accuracy and a full material mass balance to be constructed[3,4] for the nutrients, and for factors such as total energy inputs and outputs. It enables the yield of biomass on the substrate and the oxygen uptake and carbon dioxide output rates to be accurately measured and controlled, and the flow of nutrients can be carefully balanced. It is true to say that much recent process control has developed from an understanding of the nature of a nutrient-limited continuous culture of microbes. All of this is not possible in a batch culture with its very rapidly changing environmental conditions, e.g., of cell concentration, nutrient concentration, disolved oxygen tension, etc.

III. SOME PARAMETERS IMPORTANT TO SCP PRODUCTION

A. Substrate

This has been largely dealt with in the introduction and it is fairly obvious that for any large scale process (100,000 tons per annum) the substrate should be (1) cheap, (2) readily available in large amounts close to the proposed site of SCP production, (3) contain no toxic residues, (4) pose no intractable process problems, and (5) support microbial growth. For these reasons some of the favored substrates[5,6] for a large-scale process are methane (natural gas), methanol, and paraffins although ethylene, ethanol, or renewable substrates or industrial wastes might also be attractive.

B. Organism

1. Type

Algae, filamentous fungi, bacteria, and yeasts may form the basis of suitable SCP processes. In some cases as for growth on methane no yeast or fungus has been isolated which can grow on the substrate, but for methanol both bacteria and yeasts can be used. As the product is intended as a protein source the use of bacteria with their crude protein content of 60 to 80% of the biomass is favored compared to the 40 to 60% crude protein content of yeasts (the crude protein is the total nitrogen in the product multiplied by 6.25, and will include nucleic acids, cell wall compounds, etc). One advantage of using yeasts is that they are larger (5 μ diameter) than bacteria (1 to 2 μ diameter) and the costs of biomass removal from the product stream by flocculation/centrifugation are lower.[88] Other factors that have to be taken into account include the acceptibility of the product by the public; for example, yeasts are regarded by some people as a "natural" foodstuff and may be favored. However some yeasts when grown on methanol have a flavoprotein-mediated methanol oxidase system[7] which may give a very low yield on methanol compared to bacteria. Some bacteria produce intracellular poly-β-hydroxybutyric acid granules[8] under nutrient limitations other than carbon-energy limitation; this is a largely indigestible compound[9] and may lower the nutritive value of the product. The bacteria used for the methane and methanol based processes have not been reported to produce this polymer but *Methylococcus* NCIB 11083 which is the basis of a natural gas SCP process can produce an intracellular glycogen like polymer[10] which is probably readily digestible and may be nutritionally beneficial.

Other factors which influence the choice of organism are the yield of biomass on the carbon substrate and on oxygen, the maximum growth rate, the maximum and optimum growth temperatures, and the robustness of the organism in a large fermenter. Obviously, a high yield is important and will minimize oxygen consumption and heat output and maximize productivity. Growth at a high temperature (35 to 60°C) will minimize cooling costs, but too high a temperature will cause a low oxygen and methane solubility. All these factors have to be optimized in any process.

2. Enrichment

One of the initial steps in the development of a process for the production of single-cell protein from methane or methanol is the isolation of organisms capable of utilizing these substrates as their sole source of carbon and energy. The enrichment procedure should as far as possible be designed to approximate closely the conditions (i.e., pH, temperature, concentration of carbon source, nitrogen source, etc.) to be employed in the final production process. The basic enrichment procedure used by a number of workers[11-13] for the isolation of organisms capable of utilizing methane or methanol as their sole source of carbon and energy consists of enrichment in batch culture, and this may be followed by a second selection in continuous culture.

a. Batch Enrichment

If the organism sought is present in small numbers then the chance of successful enrichment may be proportional to the size of the inoculum.[14] Naturally the chances of isolation are enhanced if the organism is isolated from an environment where the culture substrate is found, e.g., in areas where free methane is found. A sample of suitable material (soil, mud, sediment, sewage, etc.) is used to inoculate minimal salts medium containing methane or methanol as the sole source of carbon and energy. Incubation is carried out at the required pH and temperature, with the nitrogen source of choice. Johnson[14] has suggested that an atmosphere of CO_2 (at least 1% v/v) is essential during the initial enrichment and during routine enrichments for methane and methanol utilizers in our laboratory this procedure was found to be successful. A problem may arise if growth is inhibited by high dissolved oxygen tensions, however once growth commences adequate concentrations of oxygen must be maintained and in practice this is achieved using well agitated shake flask cultures, with the flasks containing approximately 10 to 20% of their capacity of medium.[13,14] As soon as rapid growth occurs the enrichment culture should be transferred to fresh medium. If transfer is delayed than the enrichment will not be as selective or as rapid.[14] The reason for this is that the exhaustion of the supply of methane or methanol and the accumulation of metabolic products may result in the replacement of the predominant species by a second population which may be growing at the expense of carbon substrates other than the specific defined carbon source.[15-17] After serial transfer of the enrichment culture, attempts to isolate the desired organism should commence. However, although the existence of methane-oxidizing bacteria has been known since the turn of the present century, attempts to isolate these organisms in pure culture and to classify them were not successful until the methodology of isolation of these organisms was described by Whittenbury et al.[13] Thus, before the isolation methodology was established, enrichment culture using minimal salts medium and methane as the sole source of carbon and energy resulted in mixed cultures[11] from which the methane utilizing organism was not isolated in pure culture.

b. Continuous Enrichment

Jannasch[18] has pointed out that the complex processes occuring in a batch culture inoculated with a sample of soil are largely uncontrollable, consequently only organisms of pronounced metabolic specificity (e.g., high maximum growth rate) are successfully and reproducibly enriched for. Competition for the growth limiting substrate in the chemostat gives conditions quite dissimilar to those in a batch culture. The continuous supply of the growth limiting nutrient allows continuous growth, and the concentration of cells and products reaches a steady-state.

Although the principles of enrichment and selection in continuous culture were described in 1950[19,20] this technique has been seldom used, although it offers the possibility of selecting for microorganisms in an extremely wide range of controlled environments. In a number of cases where continuous enrichment procedures were used[16,21] the methodology was poorly described and the rationale for some of the schemes used is often difficult to ascertain. For example, Cremieux et al.[12,16] used initial enrichment in batch culture at 30 to 32°C in a medium containing yeast extract. This was followed by enrichment in continuous culture at 34°C, D = 0.15 hr^{-1}. The yeast extract was then omitted from the medium and the culture gradually adapted for growth at 40°C at a dilution rate of 0.3 hr^{-1}. It was not surprising, therefore, to find that the resulting mixed culture performed most efficiently in terms of yield from methanol and oxygen at a temperature near the initial isolation temperature of 30°C, but at 41.6°C the yield on methanol and oxygen was substantially lower. Another procedure commonly applied is the gradual increase in the dilution rate; the idea being that only the most rapidly growing organism avoid being washed out. These procedures have an important setback as once a culture has been enriched in batch culture and then grown for any length of time in continuous culture at a low growth rate one organism capable of utilizing the sole carbon and energy source (methane or methanol) will predominate at the expense of other organisms. Consequently, subsequent alteration of the culture conditions such as a gradual increase in the dilution rate or growth temperature will only result in the selection of variants of the organism initially enriched for. Although these variants may adapt to growth under the altered environmental conditions, they are not necessarily the best suited for growth in such an environment. Unless the enrichment in continuous culture is continuous (i.e., the fermentor is continuously and deliberately inoculated) the selection pressures immediately applied in the continuous culture environment must approximate closely to those required for the desired process.

These principles have been admirably demonstrated by Kuenen and co-workers.[22-24] Two chemostats were connected to a reservoir containing an appropriate medium with a growth limiting nutrient. The medium pumps were adjusted to give different dilution rates (0.03 and 0.3 hr^{-1}) in the two chemostats, while all other parameters such as temperature, pH, dissolved oxygen, etc. were maintained at the same value. Both vessels were inoculated with an identical sample of water. Under carbon limitation two completely different populations were selected at these different growth rates. At low dilution rates (0.03 hr^{-1}) organism A, with a low μ_{max} and a high affinity for the substrate was selected, while at the high dilution rate (0.3 hr^{-1}) organism B with a high μ_{max} and a relatively poor affinity for the limiting substrate was selected. When experiments were conducted with a mixture of the two isolates, A became dominant at low and B became dominant at high dilution rates.

C. Mixed Cultures or Pure Cultures for SCP Production

As pointed out previously, enrichment cultures using minimal salts medium and methane as the sole source of carbon and energy resulted in mixed cultures from which the methane utilizing organism could not be isolated in pure culture. This situation stimulated interest into the possible roles played by the various microorganisms comprising these mixed cultures. Early work[11] suggested that the components in the mixed culture utilized the potentially inhibitory methane oxidation products and thus prevented their accumulation and inhibitory effect on the methane utilizing organism. Wilkinson et al.[17] isolated a stable mixed culture which grew on methane and which contained in addition to the methane utilizing *Pseudomonas* sp a methanol utilizing *Hyphomicrobrum* sp with a low K_m for methanol and two heterotrophic bacteria (an *Acinetobacter* and a *Flavobacterium* sp) that were unable to grow on C_1 compounds. It was deduced that the inability of the *Pseudomonas* sp to grow well in pure culture with methane as the sole source of carbon and energy was due to the accumulation of methanol in the culture supernatant. In the mixed culture the accumulation of methanol was prevented by the presence of the *Hyphomicrobrum* sp which utilized this compound for growth. The role of the other two organisms was suggested to be the removal of complex biological growth or cell lysis products. Attempts to isolate the methane utilizing pseudomonad were unsuccessful.

The early success with mixed cultures capable of utilizing methane[11,21] or methanol[11] coupled to the difficulty experienced at this time in the isolation of these C_1 utilizing organisms, led some workers[25] to strongly advocate the use of mixed cultures for the production of single-cell protein. Harrison and co-workers[25,26] suggested that monocultures rarely, if ever, occur in natural environments, but are artifacts of the laboratory, and that populations with commensal relationships have evolved to suit every environment. While strongly agreeing with these sentiments we argue that the aim of the industrial microbiologist is not to simulate the natural environment but to grow organisms in the artificial environment of the laboratory or plant under conditions that are optimal for biomass or product formation, functions that may be unimportant to the organism in its natural environment. This does not mean that the use of mixed cultures is ruled out; on the contrary, there are good reasons for the use of mixed cultures.

For example the excretion of products into the culture supernatant will cause a depressed yield coefficient.[91] However, in a culture containing a single organism the complete elimination of organic carbon compounds in the culture supernatant has not been reported (Goldberg[27]). This author has pointed out that between 5 to 10% of methanol supplied to a methanol-limited culture appears as products in the culture supernatant. Dostalek and Molin[28] reported that in a methanol limited chemostat culture of *Methylomanas methanolica* the amount of organic carbon (not identified) excreted into the culture supernatant was inversely proportional to the grow rate. Similar results[29] were reported for the growth of the methane utilizing bacterium *Methylococcus* NCIB 11083. However, the productivity of the culture supernatant carbon of a culture of *Methylomanas methanolica* was inversely proportional to the dilution rate, whereas in a culture of *Methylococcus*, the productivity increased with an increase in the dilution rate.

When *Methylococcus* NCIB 11083 is grown as a pure culture in a chemostat, carbon compounds amounting to approximately 28% of the bacterial carbon are "excreted" into the culture supernatant at low growth rates. Products of methane oxidation such as methanol, formaldehyde, and formate did not constitute a significant amount of the organic carbon, which could be accounted for mainly as protein and nucleic acid. The relationship between growth rate and the concentration of protein and nucleic acid in the culture supernatant and whole cells of *Methylococcus* was similar. Moreover, the culture supernatant did not contain a single protein but a mixture of protein which on SDS-polyacrylamide gels had a similar profile to the protein extracted from whole *Methylococcus* sp cells. The authors suggested that these compounds originated from a growth-dependent lysis of the methane utilizing bacterium.[29] In a mixed culture containing *Methylococcus* NCIB 11083 plus four heterotrophic bacteria (the latter organisms being unable to grow on methane), the organic carbon in the culture supernatant was reduced to a low concentration and remained independent of growth rate.[29] The heterotrophic bacteria present in the mixed culture constituted by numbers approximately 14% of the total population and possessed between them extracellular proteases and nucleases, as well as lipases and peptidases. These extracellular enzymes were thought to play an important role in reducing the amount of carbon in the culture supernatant to a low concentration. One explanation for the presence of residual carbon in the culture supernatant of the mixed culture is that it is composed of minor cell constituents which are individually present at concentrations too low to be utilized by the heterotrophic organism, i.e., at concentrations well below their K_m value(s).

There was no significant difference between the pure culture of *Methylococcus* and the mixed culture containing this organism, in terms of yield from methane and oxygen or in μ_{max}. However, the mixed culture appeared to have a number of advantages over the pure culture. The pure culture foamed a great deal and this problem was eliminated in the mixed culture, although antifoam agents were not examined and may have eliminated this problem from the pure culture. The mixed culture appeared to be more resistant to pertubations in the culture environment, e.g., changes in temperature, pH, and ammonia concentration. The major advantage of the mixed culture is the removal of extracellular carbon which is present in the culture supernatant of this pure culture.[29,91] The presence of high concentrations of carbon in the culture supernatant could create problems in a scaled-up process due to the accumulation of organic carbon in recycled process water and/or in an effluent stream. This problem is illustrated when a pure culture of *Methylococcus* NCIB 11083 is grown at high productivity and a proportional increase in the culture supernatant carbon is observed. Although the supernatant carbon concentration in the mixed culture increased at high productivity, it remained substantially lower than that observed in a pure culture.

Unfortunately, data for the amount of carbon in culture supernatants of pure cultures of methanol or methane utilizing bacteria grown at high productivities is not available. If the data reported for *Methylomonas methanolica*[28] is representative of methanol utilizing organisms as a whole, then at a known productivity of $6\ g \cdot \ell^{-1} \cdot hr^{-1}$ ($D = 0.2\ hr^{-1}$ supernatant carbon $= 1.8\ g\ell^{-1}$) the amount of organic matter released into the culture supernatant (organic matter = 46% carbon) will be 0.8 g/ℓ/hr. This is potentially a problem if water recycle is an integral part of the process and may even necessitate the inclusion of an effluent treatment plant to reduce the level of organic carbon in the effluent discharge. We would suggest that the presence of organic compounds in the culture supernatant is a characteristic of obligate C_1 utilizing organisms, and until an organism has been isolated which overcomes this problem the use of mixed cultures will probably be necessary when processes are operated at high productivities.[29,91] One important long-term disadvantage of mixed cultures may be that in a mixed culture there may be a tendency to select for mutants of the major organism which release greater amounts of carbon into the culture and result in an increased proportion of secondary organisms.

Goldberg[27] has emphasized that the formation of any product other than biomass will cause a decrease in the yield coefficient. We endorse the view expressed by this author that evidence showing that mixed cultures are inherently better than pure cultures, in terms of yield and maximum growth rate, is poor. Harrison et al.[30] described a mixed culture containing one methanol utilizing organism, *Pseudomonas* EN, and

Table 1
SOME YIELD COEFFICIENTS FOR VARIOUS BACTERIA OR YEASTS GROWN IN CHEMOSTAT CULTURE WITH METHANOL AS THE SOLE SOURCE OF CARBON AND ENERGY AND AMMONIUM SALTS AS THE NITROGEN SOURCE

Organism(s)	Culture conditions	Carbon incorporation pathway	Y_{CH_3OH} (g bacterial dry weight mole methanol^{-1})	Y_{O_2} (g bacterial dry weight mole oxygen^{-1})	Ref.
Pseudomonas C (pure culture)	32°C, low productivity	RP[a]	17.28	—	76
MSI (mixed culture)	34°C, high productivity	RP?[a]	14.08[b] 13.1	19.52[b]	16
Methylomonas clara (pure culture)	39°C, high productivity	?	16	—	71
Methylomonas methanolica (pure culture)	35°C, low productivity	RP[a]	12.48	18.24	77
Methylomonas methanolica (pure culture)	30°C, low productivity	RP[a]	15.36[b]	16.96[b]	28
Pseudomonas 'EN' (pure culture)	42°C, low productivity; citrate in medium	?	9.6—14.7[c]	—	25,26,30,90
Pseudomonas 'EN' (mixed culture)	42°C, low productivity	?	17.28	—	25,26,30
Mixed bacterial culture	50°C, low productivity	?	13.4	—	15
Pichia methanotherm MO-104 (pure culture)	37°C—40°C, high productivity	?	12.4	—	78
Hansenula polymorpha DL-1 pure culture	37°C, high productivity	?	11.5	—	79

Note: That the bacteria were grown between pH 6.5 and 7.5 whereas the yeasts were grown at a pH of between 4.0 and 6.0; and high productivity in general refers to a productivity of over 2 $g\ell^{-1}$ hr^{-1}.

[a] RP = ribulose monophosphate pathway.
[b] Refers to a yield corrected for maintenance methanol or oxygen consumption.
[c] No citrate in medium.

four heterotrophic bacteria that were unable to utilize methanol. These authors reported that *Pseudomonas* EN performed better both in terms of yield and μ_{max} in a mixed culture than as a pure culture. It should be noted that earlier work with the pure culture of *Pseudomonas* EN was carried out in medium containing citrate as a chelating agent. When this chelator was omitted from the medium the performance of *Pseudomonas* EN in pure culture improved markedly and approached closely the yield and μ_{max} observed with the mixed culture (Table 1). As Goldberg[27] has suggested, with further optimization of growth environment the yield and μ_{max} would probably not be significantly different between the pure and mixed culture. Indeed, Goldberg has shown this to be the case in *Pseudomonas* C where addition of other nonmethanol utilizing organisms to a pure culture of this organism, grown under conditions optimal for biomass production, had no effect on yield coefficient or μ_{max}. A comparison of yield values obtained on methanol with pure or mixed cultures is shown in Table 1, and indicates clearly that although the mixed culture shows higher yield values when compared to the yield observed when the component methanol utilizer is grown in pure culture these values are not as high as those obtained when the growth of the monoculture is optimized.

When the growth substrate is a pure source of methane or methanol the need to use mixed cultures will largely depend on the culture supernatant organic carbon content at high productivities. If a pure culture can be grown at high productivities with little or no excretion of organic compound into the supernatant the process will have a number of advantages over the mixed culture situation. The most important of these will be from the standpoint of registration of the product to meet the statutatory guidelines stipulated by the regulatory authorities in most countries. The second is that the quality of the product will be easier to maintain and monitor. However, if substantial amounts of organic carbon are excreted into the culture supernatant a nutrient rich environment will ensure and it will be difficult to prevent contamination. Under these conditions it would be better to employ a defined mixed culture of the type described by Linton and Buckee[29] The mixed culture will not exhibit significant improved yields over the pure culture unless very considerable cell lysis[91] takes place, but will utilize the culture organic carbon that accumulates at high productivity and thus eliminate problems of foaming and may even reduce the level of contamination, although this latter point has not been deliberately tested.

Mixed cultures are essential for SCP production from natural gas where the natural gas has an appreciable ethane and propane content[31,54] This is because ethane is cooxidized by methane utilizing bacteria. For example *Methylococcus* NCIB 11083 can oxidize ethane to acetate, and propane to propionic acid and acetone. In a pure culture these products would accumulate and inhibit growth. For example in a high productivity (5.4 g ℓ^{-1} hr^{-1}) process when the ethane was approximately 10% of the methane concentration the extracellular acetate concentration, without any recycle of the water stream, would have been between 0.1 *M* and 0.2 *M* if a mixed culture had not been used. This concentration of acetate caused a complete inhibition of the growth of a pure culture. Another possibility would be to use a two stage system where an ethane utilizer grew on the ethane and propane in the first stage and the resultant pure methane would be used by *Methylococcus* sp in the second stage.

D. Growth Rate

Although much attention has been placed on the importance of yield of organisms from the substrate, less attention has been placed on growth rate. The growth rate is important because it may limit the process productivity (which is the product of organism concentration (g ℓ^{-1}) and specific growth rate (hr^{-1}). In a steady state continuous culture the specific growth rate equals the dilution rate (assuming 100% culture viability) and the productivity, P, is given by:

$$P = D \cdot x \ (g\ \ell^{-1}\ hr^{-1}) \tag{1}$$

where D is the dilution rate (hr^{-1}) and x is the steady-state biomass concentration. A decrease in product recovery costs is generally obtained with productivities of up to 3 to 5 g ℓ^{-1} hr^{-1} and it is often stated that no economical benefits are obtained by an increase in productivity much beyond these values.[32] This is however a rather unresearched area and although bacteria have been grown up to 50 g ℓ^{-1} in batch culture,[33] no long-term continuous process data has been published on this subject. In theory bacteria might be grown to densities approaching their packing density, but the rate of diffusion of substrates, or product production may become a problem, and an enormous power input would be required to give a high enough rate of oxygen transfer. In practice oxygen transfer will probably limit productivity with hydrocarbons as substrate although other factors, e.g., increased culture viscosity due to some cell lysis and subsequent nucleic acid accumulation, should not be disregarded. What is clear is that to operate a process at a maximum productivity of about 6 g ℓ^{-1} hr^{-1}, biomass of 30 g dry weight ℓ^{-1} requires a dilution rate of 0.2 hr^{-1}.

For a methane or methanol based process it has frequently been argued that those organisms with a predicted high yield should be used (i.e., those bacteria with the ribulose monophosphate pathway as opposed to the serine pathway of carbon assimilation).[34] An examination of some published growth rates[27,35] suggests that another reason for using an organism with the ribulose monophosphate pathway is that it can readily be grown at a dilution rate of 0.2 hr^{-1} whereas this may be impossible for the slower growing serine pathway organisms.

The growth rate has several other influences on the process. The nucleic acid content, especially the ribonucleic acid content of bacteria, increases with growth rate.[36] Although the nutritional value of nucleic acid nitrogen for animals is unknown it is well known that high dietary nucleic acid concentrations are not tolerated by humans.[37] This is probably of little consequence as most commercial processes in development will aim to produce a product for the animal feedstock industry. A further influence of growth rate is that according to the equations of Pirt[38] the faster the growth rate in carbon-energy limited culture the higher will be the observed yield because the maintenance substrate consumption will be a smaller percentage of the substrate utilized at fast, as opposed to slow, growth rates. This point was fully appreciated by Abbott and Clamen[39] in their theoretical study of the economics of SCP production from a wide variety of substrates. The formulation of Pirt can be written as:

$$\frac{1}{\text{Y (sub)}} = \frac{1}{\text{Y (sub) theoretical}} + \frac{\text{M (sub)}}{\text{D}} \qquad (2)$$

where M (sub) is the maintenance substrate consumption (mol g^{-1} hr^{-1}) and Y (sub) is the observed yield (g mol^{-1}) while Y (sub) theoretical represents the yield at an infinite growth rate and represents the yield in the absence of any maintenance substrate consumption. Naturally, in any industrial process it is the observed and not the theoretical yield which is of importance, but the faster the growth rate the higher will be the observed yield. A further implication of growth rate is that for most cultures the respiration rate increases with growth rate according to the equation of Harrison and Loveless.[40]

$$QO_2 = \frac{D}{YO_2 \text{ (theoretical)}} + MO_2 \qquad (3)$$

where QO_2 is the *in situ* respiratory activity of the growing culture (mol O_2 hr^{-1} g^{-1}, MO_2 is the maintenance oxygen consumption rate (mol hr^{-1} g^{-1}), and YO_2 (theoretical) represents the yield on oxygen in the absence of any maintenance oxygen consumption. Thus as the growth rate is increased the *in situ* respiration rate will increase and the power input required will have to be increased to maintain adequate aeration at a high cell density. This is especially true for microbial growth on reduced substrates like methanol and methane where there is a very low yield on oxygen, i.e., a high oxygen requirement for growth.[41,42] This high oxygen requirement means that oxygen transfer will often limit productivity. Naturally for a gaseous carbon-energy substrate such as methane the same argument applies and an increased rate of methane gas transfer is

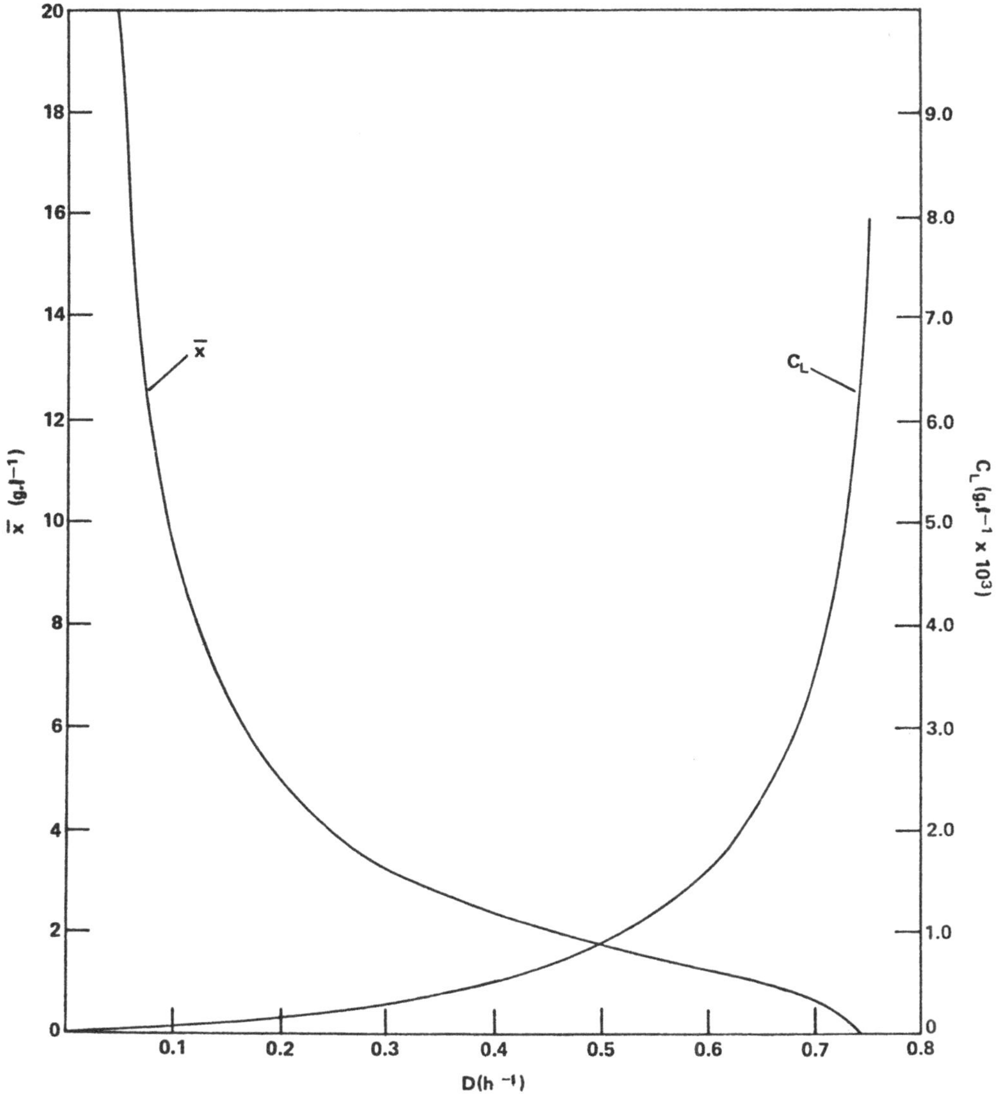

FIGURE 1. Theoretical relationship between steady-state biomass and dilution rate for a gas (oxygen)-limited chemostat culture.[43] A very similar relationship holds for a methane-limited culture.[84] Note that for a methane-grown culture it may be difficult to determine if the culture is methane- or oxygen-limited because an increase in the concentration of either component in the gas phase will not necessarily cause an increase in the steady-state biomass.[84,83] $\overline{X}$ represents the steady-state biomass, C_L the dissolved oxygen concentration when $\mu m = 0.8$ hr^{-1}, Ks for oxygen $= 0.5 \times 10^3$ gℓ^{-1}, $YO_2 = 1.25$ g g^{-1}, Kla $= 100$ hr^{-1}, and Cs the saturation concentration of oxygen $= 0.8 \times 10^2$ gℓ^{-1}. (From Harrison, D. E. F., *J. Appl. Chem. Biotechnol.*, 22, 1—4, 1972. With permission.)

required to maintain productivity as the dilution rate is increased. This system was modeled by Topiwala.[43] Figure 1 shows the relationship between steady state biomass and dilution rate for a gas-limited (i.e., oxygen or methane-limited), continuous culture. It is obvious that as the dilution rate is increased a higher rate of gas transfer will be required to avoid the dramatic decrease in steady-state biomass shown in Figure 1. It is evident that if the rate of gas transfer is constant, yet the specific rate of gas uptake by the microbes increases with dilution rate, then there will be a decrease in steady-state biomass when the dilution rate is increased in a gas-limited culture.

E. Affinity of the Organism for the Substrate

The affinity of the microorganisms for the growth limiting substrate is important for several reasons. For a liquid carbon-energy source such as methanol it is desirable to run the continuous process under conditions of carbon-substrate limitation. The reasons for this are: (1) under conditions of carbon-energy limitation the protein content of the biomass is maximal, i.e., no nonnitrogenous storage products are likely to be formed in large amounts and there will be no "overflow" of intermediate metabolites[44,45] into the culture supernatant; (2) it is often stated that the yield of microorganisms are highest under conditions of carbon-substrate limitation, although as yet there is little published evidence for this in the methylotrophs; and (3) it is highly desirable to maintain as low as possible a concentration of potentially toxic substrates such as methanol in the fermenter. The substrates at high concentrations may, as is the case of methanol, be toxic to the growth of the microorganisms themselves,[46] or may accumulate in the product. This problem can be heightened when recycle of some of the effluent liquid stream (minus biomass) into the fermenter is envisaged.[47] Under conditions of recycle the concentration of the unused substrate may increase and for a substrate such as methanol, growth inhibition or a drop in yield may occur as the oxidation of excess methanol is "uncoupled" from biomass formation.[46,48]

In the case of methane as a substrate there is a further reason for operation of the process under methane limitation and that is because methane is supplied as a gas which, like oxygen, has a relatively low solubility in culture media (approximately 1062 μM for a saturated solution at 40°C) and there is the problem of maintaining a high methane mass transfer rate to support a high steady-state biomass. In a manner analagous to that for oxygen transfer, the rate of methane transfer (mol ℓ^{-1} hr^{-1}) can be given by:

$$\text{Rate of methane transfer} = \text{Kla (Tg} - \text{T1)} \qquad (4)$$

where Kla represents the overall mass transfer coefficient for methane, Tg is the methane tension in the gas phase which can be expressed as the saturation concentration of methane in a solution which is in equilibrium with the given fermenter head space methane concentration (mol ℓ^{-1}), and T1 is the dissolved methane concentration (mol ℓ^{-1}. Obviously, one way to maintain a high rate of methane transfer is to maintain a low value for T1, i.e., to operate methane limited when the dissolved methane concentration will be low and the driving force for methane transfer will consequently be high. Under methane limitation the dissolved methane concentration will be governed by the Ks of the bacteria for methane and the dilution rate. The Ks represents the dissolved substrate concentration which will support half of the maximum specific growth rate (Figure 2) and can be related to the specific growth rate and substrate concentration by the well-known "Monod" equation[49,50]

$$\mu = \mu_{max} \frac{\tilde{s}}{Ks + \tilde{s}} \qquad (5)$$

where μ is the specific growth rate (hr^{-1}), $\tilde{s}$ is the dissolved substrate concentration (mol ℓ^{-1}), and Ks is the substrate concentration (mol ℓ^{-1}) which supports half of the maximum specific growth rate, μ_{max}. For *Methylococcus* the value of the Ks for methane has been estimated from oxygen electrode experiments as being approximately 30 μM,[29] and for methanol-grown *Pseudomonas extorquens* was 20 μM,[51] both these values are of the same magnitude as the Ks of many heterotrophic bacteria for sugars.[52] However, in the case of sugars the Ks relates to a specific transport mechanism whereas for methane and methanol the substrates probably enter by simple diffusion and the

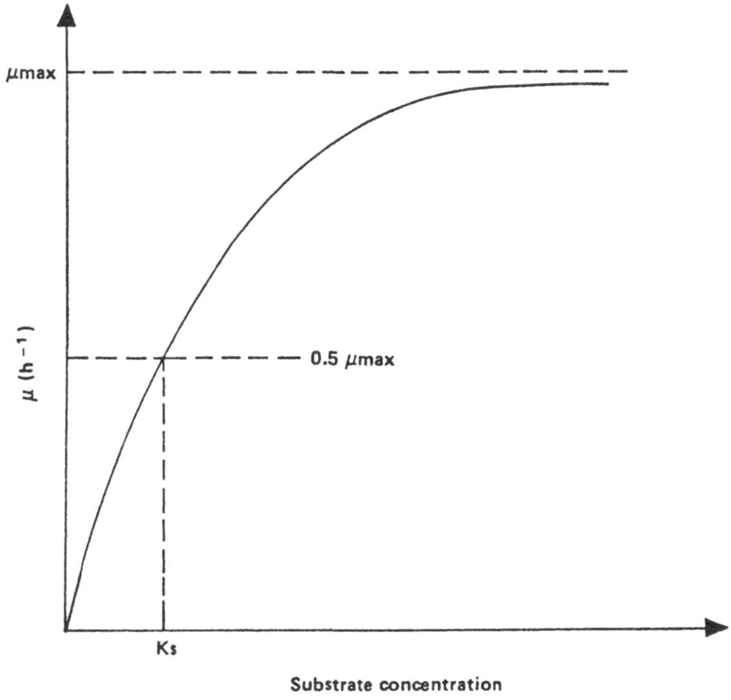

FIGURE 2. Theoretical relationship between the specific growth rate μ (h^{-1}) of a nutrient-limited microbial culture and the nutrient concentration. For a nutrient-limited chemostat culture the specific growth rate equals the dilution rate (equivalent to the inlet liquid flow rate divided by the liquid volume in the fermenter) and the steady-state, growth-limiting nutrient concentration usually refers to the steady-state concentration in the culture supernatant. Ks represents the growth-limiting nutrient concentration which will support half the maximum specific growth rate.

Ks relates to the Km of the intracellular enzymes for methane and methanol. Obviously there are other ways to increase the overall rate of methane transfer, e.g., by an increase in the value of Tg or Kla in Equation 4 but a low concentration of dissolved methane will help to maximize the rate of methane transfer.

There is one possible disadvantage of operation under carbon-substrate limitation where there is a very poor affinity, i.e., high Ks for the substrate. Under these conditions the maximum dilution rate under which the culture can be operated may be well below the maximum specific growth rate. This is because the dilution rate at which "wash-out" of the culture occurs, the critical dilution rate Dc (hr^{-1}) is given by:[50,52]

$$Dc = \frac{\mu m \, Sr}{(Ks + Sr)} \qquad (6)$$

where Sr is the growth limiting substrate concentration (mol ℓ^{-1}) in the influent medium, Ks is the substrate concentration (moles ℓ^{-1}) which supports half the maximum specific growth rate, μm (hr^{-1}). Hence with a high value of Ks for the substrate the dilution rate, Dc, will be considerably less than the maximum specific growth rate, μ_{max}. In practice this is extremely unlikely to be a problem in any industrial situation except for the possible production of SCP from natural gas with a considerable ethane content. This is because ethane is a competitive[53] inhibitor of methane oxidation and will effectively increase the Ks of the cultures for methane and consequently decrease

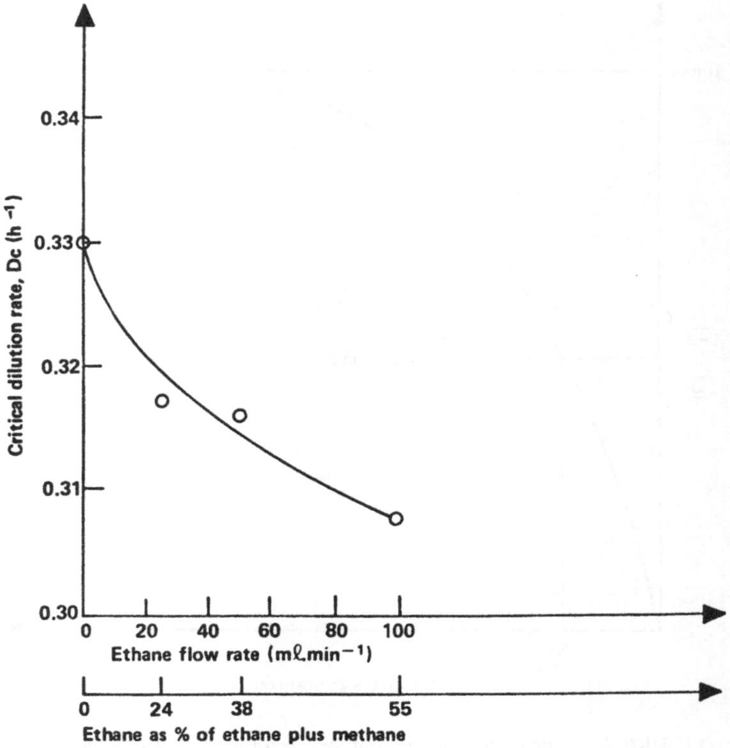

FIGURE 3. Observed relationship between the concentration of ethane in the inlet gas phase and the critical dilution rate of a methane-limited mixed bacterial culture which contained *Methylococcus* NCIB 11083 and heterotrophic bacteria. The culture was grown at productivities of up to 0.7 $g\ell^{-1}$ hr^{-1}, 45°C, pH 6.6, air flow 500 $m\ell/min^{-1}$, and methane flow 91 $m\ell/min^{-1}$. The heterotrophic bacteria grew on products released by the lysis of some of the *Methylococcus* cells,[29] and on acetate produced from the cooxidation or ethane by the *Methylococcus*[13,31,53,80,85,86] which grows on methane. At no stage was free acetate detected in the culture. A pure culture of the *Methylococcus* could not be grown under such conditions because acetate accumulated and inhibited growth.

the value (see Equation 6) of the critical dilution rate. For *Methylococcus* NCIB 11083 the Km values for methane and ethane oxidation are both approximately 30 μM [31,54] and both are likely to be competitive inhibitors of each other.[53] It can be easily calculated[31] that the presence of ethane in natural gas will under most conditions only give a slight decrease in the critical dilution rate of a methane-limited culture and will not limit the process productivity. Figure 3 shows the variation in critical dilution rate with ethane concentration for a methane-limited mixed bacterial culture containing *Methylococcus* NCIB 11083. The presence of ethane has only caused a slight decrease in the critical dilution rate. It can be shown that even when the total gas pressure is increased there will only be a slight decrease in the critical dilution rate.

As stated earlier the high oxygen requirement for growth on hydrocarbons means that many processes may be run under conditions which approach oxygen limitation. Operation under oxygen limitation with excess carbon source would maximize the rate of oxygen transfer but could be associated with intracellular product formation and a low biomass protein content. The Ks of bacterial cultures for oxygen is very low, probably less than 1 μM, and in bacteria has only recently been accurately measured.[55] Thus a process can be operated at very low oxygen tensions without being oxygen limited.

F. Yield

The yield of bacteria on the carbon-energy substrate and on oxygen are very important parameters in the overall economics of SCP production. However it should be considered in conjunction with other factors, e.g., growth rate. The importance of yield can be related to the following:

1. The productivity of a process under carbon-energy substrate limitation will be related to the yield of organisms on the carbon-energy substrate. This will ultimately relate to the plant size required for a given productivity and will thus influence the capital costs.
2. The higher the yield on a specific carbon-energy substrate the higher will be the yield on oxygen[42,56] (Figure 4), and the less will be the power input required to avoid oxygen limitation at a given productivity. This is especially true for the highly reduced substrates such as methane and methanol where there is a high oxygen demand for growth (Table 2), and one of the major cost factors is to provide enough oxygen to maintain high productivity and avoid possible oxygen limitation.
3. On the assumption that the only products of growth are cells, CO_2, and water, then for a given substrate the lower the yield the greater is the amount of energy in the substrate which is not used for biomass formation and the greater is the heat output (Figure 5).[56]

The heat of combustion of microbial cells falls within the range[39,56] of 4.7 to 5.4 kcal gram cells^{-1} and the heat of combustion of the substrate can be obtained from suitable tables[57] For highly reduced substrates such as methane or methanol the heat of combustion is high[57] and to minimize cooling costs it is important to maintain as high a yield on the carbon-energy substrate as possible. It must be remembered that this is only an approximation because the heat of combustion for a substrate represents the maximum energy obtained whereas biological systems are not in equilibrium and the energy available may be less. If the yield of cells on the substrate is limited by the amount of ATP generated from the substrate, i.e., the culture is energy limited, then it is important to maximize the amount of ATP generated from the substrate and to minimize the amount of chemical energy in the substrate which is lost as heat. Several economic calculations have been done which correlate heat output with cooling costs,[58] but little data is available from operational pilot plants.

1. Factors Affecting Yield

If we consider factors which affect yield, then for a given organism some obvious parameters are (1) The biochemical pathways of substrate utilization, and (2) the growth conditions, e.g., nitrogen source, temperature, and dilution rate. In any discussion on yield one factor which is often overlooked is the role of the yield coefficient in microbial competition in the natural environment from which all process organisms are originally isolated. It would seem that high yield is of only minor ecological advantage to a microorganism in relation to factors such as Ks for substrate. This is best exemplified if one considers what selection pressures can be applied to a mixed bacterial population in continuous culture in order to select and favor those organisms with a high yield. To us there would seem to be no obvious selection pressure that the experimenter can apply, whereas it is easy to select for those organisms with a low Ks for the substrate, or for thermotolerance, etc. Thus it may be that in the natural environment, much to the chagrin of the industrialist, there has probably been little obvious selection pressure to favor organisms with a very high yield on the carbon-energy substrate. Further, until the biochemical factors which govern yield are fully under-

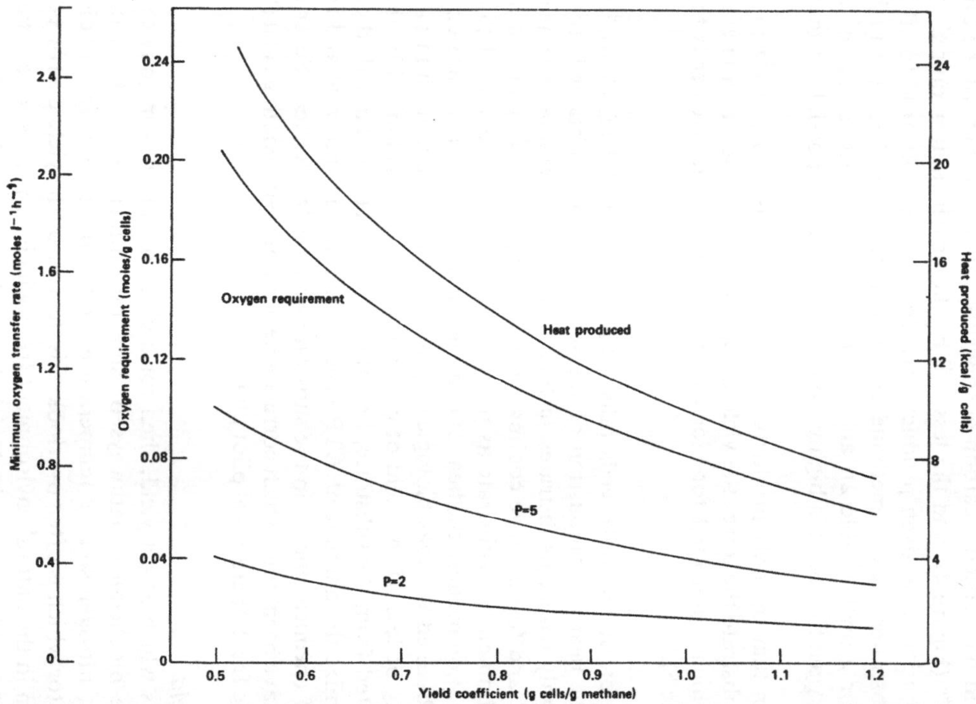

FIGURE 4. Computed relationship for a methane grown culture of yield coefficient on methane to oxygen requirement, heat production and the minimum oxygen transfer rate required at productivities of 2 g l^{-1} h^{-1} and 5 g ℓ^{-1} hr^{-1}.[83] The data for heat output was calculated from the corrlation of Cooney et al.[56] Similar relationships can be computed for microbial growth on methanol.[42] Naturally, for growth on methanol which is less reduced than methane, there will be a corresponding lower oxygen transfer rate required to maintain a given productivity[42] and a corresponding lower heat output. It should be noted that these calculations are theoretical and assume equations for microbial growth on the substrates in which the only products are cell mass, carbon dioxide, and water.[42,83] Nevertheless, these calculations are important because aeration and cooling represent two expensive items in the SCP production process. (From Barnes, C. J., Drozd, J. W., Harrison, D. E. F., and Hamer, G., *Symp. Microbioliol Production and Utilization of Gases (H_2, CH_4, CO)*, Schlegel, H. G., Gottschalk, G., and Pfennig, N., Ed., Akademic der Wissenschaften zu Göttingen, 1976, 305. With permission.)

Table 2
YIELDS ON METHANE AND OXYGEN FOR SOME BACTERIA GROWN IN CHEMOSTAT CULTURE ON METHANE

Nitrogen source	Organism and growth conditions	YCH_4 (g bacterial dry wt mole methane^{-1})	YO_2 (g bacterial dry wt mole oxygen^{-1})	Carbon incorporation pathway	Carbon balance	Ref.
NH_4^+	*Methylococcus* NCIB 11083 mixed culture (45°C, high productivity)	13.3	8.6 12.7[1] 24.3[2]	RP[3]	Yes	35
NO_3^-	*Methylococcus* NCIB 11083 mixed culture (45°C, high productivity)	10.6	7.6	RP[3]	Yes	35
N_2	*Methylococcus* NCIB 11083 mixed culture (45°C, high productivity)	8.5	5.5	RP[3]	Yes	35
NO_3^-	Mixed culture (45°C, high productivity)	9.9	6.9	?	Yes	21
NH_4^+	*Methylosinus trichosporium* OB 3b, mixed culture, 30°C high productivity	10.1	6.6	S[4]	Yes	35
NH_4^+	TM10 mixed culture, probably contains *Pseudomonas methanica* 31°C low productivity	14.4	8.32	RP[3]	No	80
NH_4^+	*Pseudomonas* sp in mixed culture 32°C low productivity	12.8 (CH_4-limited) 15.8 (O_2-limited)	7.7 (CH_4-limited) 12.8 (O_2-limited)	?	Yes	17
NH_4^+	*Methylococcus capsulatus* 37°C low productivity	16.2 (CH_4-limited) 5.0 (O_2-limited)	9.3 (CH_4-limited) 16.3 (O_2-limited)	RP[3]	Yes	81
NH_4^+	Two strains of noncapsulate bacteria, 30°C low productivity	16.0 to 17.6	—	?	Yes	13
NH_4^+	Mixed culture 33°C, high productivity	24.6[5] 16.0	23.4[5] 12.3	?	No	82

Note: Yields on methane are difficult to measure because of possible inaccuracies in the measurement of the methane gas flow rate and the amount of gas taken up.[83] In only a few cases are full carbon balances given, if the total outlet carbon from the fermenter sums up to 100% of the total inlet carbon this is a good check on the accuracy of the yield determination. In cases where there is no carbon balance it is difficult to know what was the accuracy of the yield determinations.

[1] Corrected for oxygen consumed in the methane mono-oxygenase reaction with the overall stoichiometry of $CH_4 + \frac{1}{2}O_2$.
[2] Corrected for $CH_4 + O_2$ in the methane mono-oxygenase reaction.
[4] Serine.
[3] Ribulose monophosphate.
[5] Represents yields corrected for maintenance substrate consumption.

stood, there is little in the way of genetic manipulation that can be done to increase yield.

a. Biochemistry of Methane and Methanol Utilization

If we consider methane and methanol as substrates for SCP production, then theoretically those organisms with a high yield should have the ribulose monophosphate pathway of carbon incorporation as opposed to the serine pathway.[34] Anthony[59] has argued that the ribulose monophosphate pathway organisms grown with ammonia as nitrogen source are energy limited whereas the serine pathway organisms are reductant limited as a consequence of the high reductant requirement for carbon dioxide fixation. However a complicating factor is the presence of a complete Calvin cycle for CO_2^-

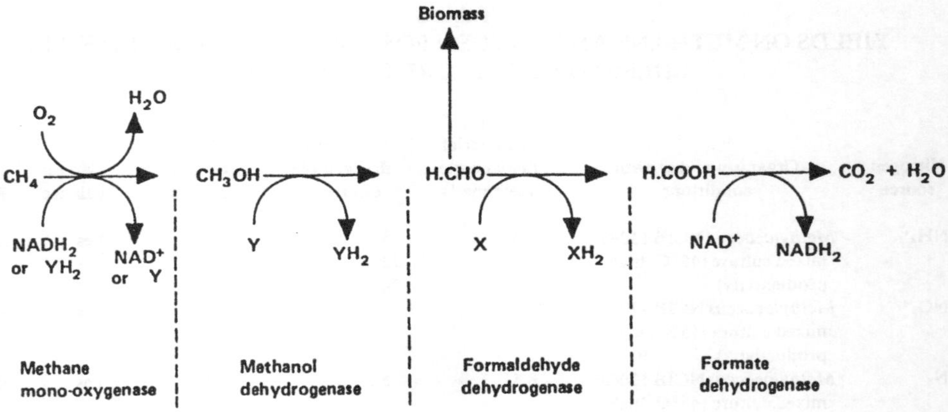

FIGURE 5. Generalized methane oxidation scheme for *Methylococcus capsulatus*[63] and *Methylosinus trichosporium* OB 3b.[62] In *M. capsulatus* the reductant for the methane monooxygenase is $NADH_2$, whereas in *M. trichosporium* it is a reduced form of a carbon-monoxide binding cytochrome *c* (YH_2) generated at the level of methanol dehydrogenase and possibly formaldehyde dehydrogenase (XH_2). In *Methylococcus* NCIB 11083 both formaldehyde and formate dehydrogenases appear to be $NADH_2$-linked (unpublished observations) whereas in many methylotrophs only formate dehydrogenase is $NADH_2$-linked[87] and the methane monooxygenase is $NADH_2$ dependent.

assimilation in the ribulose monophosphate pathway organism *Methylococcus capsulatus.*[60] The reductant dependent assimilation of an appreciable amount of CO_2 would cause a drop in yield but at the moment it is not known how much CO_2 is assimilated *in situ* by this route. A problem in many analyses is that the assumption is often made that the cultures are energy limited and not carbon or reductant limited. Although Anthony[59] has argued that the ribulose monophosphate organisms are energy limited the evidence is not conclusive. For example if quite a large amount of CO_2 is fixed then reductant limitation could occur or alternatively there could be some control mechanism at the level of formaldehyde dehydrogenase (Figure 5) such that the cultures are in fact intrinsically carbon limited. A recent correlation of growth yield data with heats of combustion[61] would indicate that the growth of organisms on reduced hydrocarbons like methane, methanol, and ethane is potentially not energy limited but may be limited by some other factor. There would appear to be a low yield on methane and methanol for the energy potentially available in these substrates, and hence, a large amount of this energy will appear as heat.

Growth on methane presents a special problem in that the methane molecule must first be hydroxylated in a reaction:

$$CH_4 + O_2 + XH_2 \longrightarrow CH_3OH + X + H_2O$$

In the serine pathway organism *Methylosinus trichosporium* OB 3b, XH_2 has been equated[62] with a reduced form of a carbon monoxide binding cytochrome *c*, whereas in the ribulose monophosphate pathway organism *Methylococcus capsulatus* (Bath) the cofactor X is $NADH_2$.[63] In *Methylosinus trichosporium* reduced cytochrome *c* is generated by methanol dehydrogenase, whereas in *Methylococcus* NCIB 11083, $NADH_2$ is generated at the level of formaldehyde and formate dehydrogenases. The observed growth yields obtained with *Methylococcus* NCIB 11083 grown on ammonia[35] would indicate that enough carbon is oxidized via formaldehyde and formate to CO_2 to almost exactly satisfy the $NADH_2$ requirement of the methane mono-oxygenase reaction and as in this organism the formaldehyde and formate dehydrogenases are NAD^+-linked, whereas the methanol dehydrogenase is not NAD^+-linked; the *Methylo-*

coccus would in fact be reductant limited and the only ATP available for biosynthesis would be that generated at the level of methanol dehydrogenase. In *Methylosinus trichosporium* the situation is somewhat different and there is a cyclic pathway[62] for reduced cytochrome *c* regeneration.

Recent comparisons of the growth yield of *Methylococcus* NCIB 11083 on methane[35] and methanol indicate that the molar growth yields on these substrates[64] are nearly identical. For an energy-limited culture this would imply that the biochemical conversion of methane to methanol is overall energetically neutral such that ATP would be generated whether a conventional cytochrome oxidase or the methane monoxygenase served as oxidant for the reductant utilized in the mono-oxygenase reaction. Obviously if the cultures are not energy limited and are reductant or carbon limited then such arguments cannot be used. It may be that there are energy "spillings" reactions in methylotrophs similar to those found in some heterotrophic bacteria.[44,45]

What can be deduced from the yield data is that overall the stoichiometry of the methane mono-oxygenase reaction can probably be written as:

$$CH_4 + \tfrac{1}{2}O_2 \longrightarrow CH_3OH$$

because the yields on oxygen minus this oxygen consumption in the methane mono-oxygenase reaction are very similar to growth yields on oxygen for methanol grown bacteria (Table 2). At first sight this is surprising because 1 mol of oxygen is used per mole of methane in the methane mono-oxygenase reaction itself. However one has to remember that when the reductant ($NADH_2$ or reduced C_{co}) is used in the mono-oxygenase reaction there is not the uptake of ½ mol of oxygen when 1 mol of reductant is oxidized via the cytochrome oxidase reaction. Obviously the fact that the molar growth yields on methane and methanol are the same means that there is a higher heat output and a higher oxygen requirement for growth on methane than on methanol. The energy available in the oxidation of methane to methanol (39.9 kcal mol^{-1}) is not conserved with the ultimate production of biomass, and will be liberated as heat.[31]

b. Growth Physiology

Although we have stated that there is not much the experimenter can do to increase the yield of a given organism on a substrate, this is not quite true. For example for energy and reductant limited cultures the maximum yield will be obtained with $NH4^+$ as nitrogen source (Table 2), rather than with NO_3^- or N_2. With nitrate there is the requirement for reductant to convert NO_3^- to NH_4^+, while in the case of dinitrogen fixation there is a reductant and ATP requirement, a value of 29 mol of ATP per mole N_2 fixed has been calculated from growth yield data in *Methylococcus* NCIB 11083.[35] Thus in terms of nitrogen supply the use of ammonia as a nitrogen source is favored although under certain circumstances a natural gas or methanol based process operating completely or with partial dinitrogen fixation could be envisaged.

Another factor which may well influence yield is the efficiency of mixing within the fermenter, especially with relation to the carbon-energy supply. When yields are measured in a small fermenter there is often near perfect mixing of all the substrates, although as Brooks and Meers[46] noted for a methanol limited culture of *Pseudomonas* the drop-wise addition of the growth-limiting substrate gave an oscillation in the methanol concentration. This is a fact not appreciated by most people who work with small chemostats under conditions of carbon-energy limitation. As excess methanol is oxidized without any biomass formation this results in a decrease in yield as the medium inflow drop size is increased at constant dilution rate. This situation was modeled by Harrison and Topiwala[65] but the exact biochemical mechanisms by which the added excess carbon-energy substrate is oxidized to CO_2 in a carbon-energy limited culture remains to be elucidated. In a large industrial fermenter operated under methanol limitation at high productivities it will be very important to maximize the mixing of the methanol, e.g., by the introduction of multiple methanol addition ports and to avoid any local build-up in methanol concentration (or any local starvation of methanol) and concomitant decrease in yield. For example, with a predicted biomass of 30g dry-weight ℓ^{-1} and a yield of methanol of 14g mole^{-1}, the equivalent methanol concentration in a complete medium needed to support this biomass would be 2.14 *M* methanol! Growth would be quite impossible if this was the initial concentration. Methanol is added as a separate stream, and for start up the concentration can be arithmetically increased such that growth is always methanol limited. With regard to methane as a substrate, the situation is somewhat better, first because the limited solubility of methane in the liquid phase means that it cannot reach locally high concentrations in the fermenter and second because the rate of methane oxidation seems well regulated[35] in *Methylococcus* NCIB 11083 and excess methane is not oxidized at a rate any faster than the *in situ* rate in a methane limited culture. Another point of interest is that on a small scale because methane is added with the gas stream there is no imposed oscillation in the dissolved methane concentration, although there may well be gradients in the dissolved methane concentration. However the methane-oxidizing bacteria pose another problem because the methane mono-oxygenase enzyme complex can oxidize ammonia[13,66] to nitrite and nitrate. This reaction not only competitively inhibits methane oxidation, but the product of the reaction NO_2^- is a toxic species and will inhibit growth. It is essential to maintain an ammonia concentration below 1m*M* which will minimize the rate of ammonia oxidation because the Km for ammonia oxidation at pH 6.5 is about 3 m*M*.[66] In a large fermenter it will be important to maintain a low supernatant ammonia concentration and to have multiple ammonia addition points in order to avoid any localized build-up in the ammonia concentration.

G. Nutrient Requirements — Problems of Supply at High Productivity

The use of continuous culture has enabled a very exact measurement of the nutrient requirements for bacterial growth to be made. The elegant studies of Tempest[67] have shown that when the growth of a bacterial culture is limited by a particular nutrient, e.g., K^+, NH_4^+, Mg^{++}, PO_4^{--}, then the bacterial cells may contain the minimum amount of that element necessary for growth under the conditions prevailing in the chemostat. The minimum requirement for a nutrient does not necessarily mean that this is the amount for optimal biomass production because:

1. The studies of Light[68] with Fe^{++} or SO_4^{--} limited yeast growth indicate that under these conditions there is a decrease in the P/O ratio from 3 to 2 via a loss of oxidative phosphorylation at "Site 1" with a consequent decrease in yield. Obviously although the medium contains the minimum iron or sulphate concentration necessary for growth such a decrease in yield in an SCP process is to be avoided.
2. If a nutrient is growth limiting then there is often the expenditure of energy to transport that nutrient across the cell membrane against its concentration gradient; this expenditure of energy may give a slight decrease in yield of an energy-limited culture. In the case of the lipid soluble substrates methane and methanol, there is probably no transport system and hence no utilization of energy for the transport of these compounds. In the case of inorganic ion limitation there may be a specific energy dependent transport system or incorporation system, e.g., the ATP-dependent glutamine synthetase/glutamate synthase route of ammonia assimilation. This is constitutive in *Methylococcus* and for an energy-limited culture would give a 5 to 7% decrease in yield when compared to the ATP-independent glutamate dehydrogenase route,[92] which has a higher k_m for ammonia.

3. For a high productivity single-cell protein process there is a requirement for a high salts concentration to support the high cell densities (20 to 40 $g\ell^{-1}$). In an industrial process there will be separate water, concentrated salts, and ammonia-gas-feed streams to the fermenter and the rate of flow of each stream will be adjusted to the requirements of the microbial population. If the salts stream was made up at pH 7.0 to a concentration necessary to support 20 $g\ell^{-1}$ biomass there would be an immediate precipitate of insoluble calcium magnesium phosphate. The concentrated salts stream is likely to be at a low pH value which will retain all the components in solution at this pH value, but it may be that when the salts stream is added to the fermenter (which in the case of a bacterial process will be at pH 6.0 to 7.0, but will be in the range of pH 4.0 to 6.0 for a yeast process) there may be an immediate precipitate of some of the calcium magnesium phosphates such that the growth rate becomes limited by the re-solution rates of these elements. Such a precipitate can also complex out trace elements such as iron or copper and make a determination of the growth rate limiting nutrient very complex.

However, bearing these points in mind the medium can be formulated with a knowledge of the minimal nutrient requirements of the culture. The work of Tempest[67] clearly showed that this minimum nutrient requirement could vary with growth parameters, especially with the growth rate. For example in bacteria there is a higher Mg^{++} requirement at high growth rates than at low growth rates; this can be related to the increase in ribonucleic content of the bacteria at high growth rates.[36,67] Such a change in nutrient requirement with growth rate or other parameter, e.g., pH, should always be borne in mind when designing a medium and is a criticism which can be applied to the pulse technique[69] of determining nutrient requirements.

In practice an industrial process will probably be operated under carbon-substrate limitation with a very slight steady-state excess of other nutrients. There will be the recycle of a large percentage of the water stream to the fermenter after the microbial cells have been separated by a flocculation/centrifugation step. When commercial grade chemicals are used which may contain toxic impurities there will be a build-up of any toxic species when recycle is initiated and subsequent inhibition of microbial growth. Topiwala and Khosrovi[47] modeled this system and showed that when there is an nonmetabolized toxic species in the salts flow, the concentration of this species in the fermenter will increase when recycle is initiated by a factor $(1 + R)$ where R is the recycle ratio defined as the ratio of recycle stream flow rate (q) to the supply stream flow rate (Q); the total feed to the fermenter is then $Q + q$. Naturally when recycle is initiated there will likewise be an increase in concentration of any exometabolite produced by the culture.

Finally it might be mentioned that the demonstration of a nutrient by chemical analysis of the culture supernatant does not mean that the nutrient is present in an excess of a growth rate-limiting concentration. The chemical and physical analysis will detect nutrients present as microprecipitates or present in small membrane fragments, etc. produced by cell lysis and such nutrients may not be available to the microbes.

H. Fermenter Design

It is not the object of this review to deal in detail with the engineering aspects of large-scale fermenter design. Much work has gone into the design of fermenters which will give the high rates of oxygen transfer with minimum power input required for a high productivity process. For a 100,000 ton per annum process, fermenters of up to or greater than 1000 m^3 are envisaged and these will be operated on a continuous basis. Although typical stirred tank fermenters have been used for SCP production, there has also been the introduction of novel fermenter types, e.g., the ICI pressure cycle fermenter.[70] An important aspect of the pressure cycle fermenter is that there are no moving parts within the fermenter itself and hence there is less chance of contamination via stirrer shaft seals. An important but often neglected factor in fermenter design is the effect that the fermenter configuration may have on yield. A fermenter designed solely to give a very high rate of gas transfer for a given power input[71] may not give optimal bulk nutrient mixing and there may be periods when the organisms are starved of the growth rate limiting carbon-energy substrate, and other periods when there is a sudden excess of the carbon-energy substrate. As previously discussed, such oscillations can cause a decrease in the yield, so in the scale-up of any process it is important not only to know the physical scale-up factors but also to know how the physiology of the organisms may be affected by the physical conditions which will be characteristic of the fermenter used. One advantage that scale-up does have is the contribution that microbial growth on surfaces within the fermenter has on the overall process is minimized because the wettable surface area to volume ratio decreases as the fermenter size is increased. However as discussed in detail by Hamer,[72] the formation of microbial films on the internal cooling surfaces (e.g., on plate heat exchanges) may adversely affect heat transfer and lead to a diminished efficiency of cooling. Hamer[73] has also pointed out that a greatly neglected area of research is the physiology of microbial growth on the surfaces and within the headspace of a vigorously agitated fermenter; organisms in droplets will be exposed to physical conditions far removed from those found in the bulk liquid phase. Little is known about the effects that, for example, the very high surface tension forces found in liquid micro-droplets may have on the physiology of the entrained microbes. Thus, in addition to the bulk liquid phase and the very large internal surfaces there is a third "micro-environment", namely the fermenter head space.

It is important to remember that the physical conditions in a large industrial fermenter operating at a high process productivity are quite different from the conditions found in a small low productivity laboratory fermenter. For example, in a large tower fermenter the hydrostatic pressure at the bottom of the fermenter near the air inlet may be between 5 atm and 10 atm, i.e., with air as the gas phase the partial pressure of oxygen in the gas phase could be up to 2 atm. If mixing is poor and there is inadequate process control one could envisage conditions where the dissolved oxygen concentration increases far above the air saturation (0.21 atm) value. It is important to know what such an increase in dissolved oxygen will have on process performance. In our laboratory Calvert and Hamer[89] noted that at high dissolved oxygen concentrations the yield of *Pseudomonas extorquens* in a methanol-limited culture halved and the rate of CO_2 output dramatically increased. Naturally, under such high productivity, high pressure conditions, the dissolved CO_2 concentration will also be very high and could itself change cellular metabolism or inhibit growth. In a high productivity process, because of the very high bacterial concentration, parameters such as dissolved substrate concentrations and pH can change extremely rapidly and it is very important to monitor and control such parameters; this problem can be compounded when there is a cycle of the culture around the fermenter, e.g., in the I.C.I. pressure cycle configuration, and extremely good monitoring and control system are required. All of these points must be appreciated by the design engineers who should always consult the microbial physiologist. Finally, considerations such as sterilization, process start-up, and process run-down have also to be taken into account in the fermenter design.

IV. CONCLUSIONS

Apart from the effluent industry the production of single-cell protein probably rep-

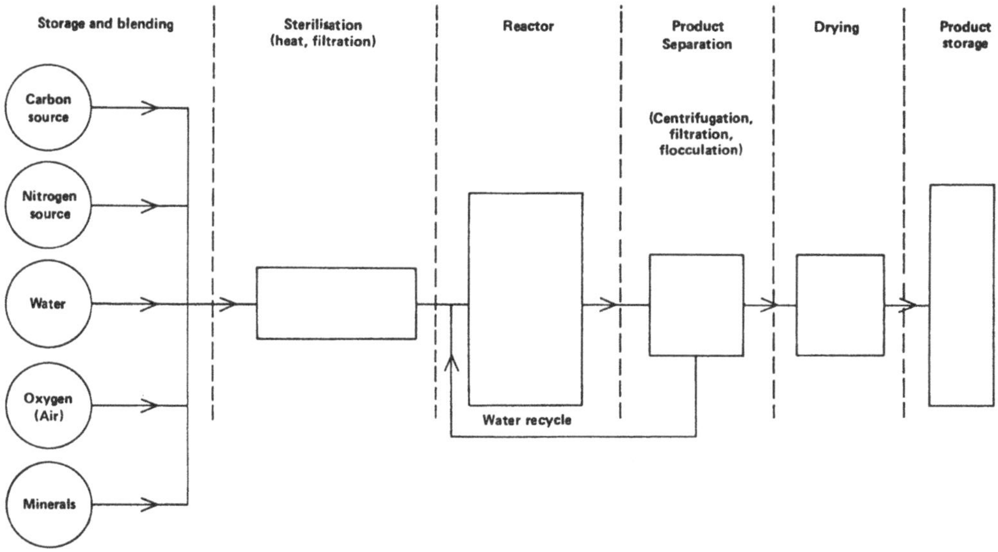

FIGURE 6. A generalized flow diagram for a typical SCP plant which shows the various unit operations. (From Topiwala, H. H., *Microbial Growth on C_1-Compounds,* The Society for Fermentation Technology, Osaka, Japan, 1975, 199. With permission)

resents the first major use of continuous culture for an industrial process. The relatively low value of the product compared to a fine chemical and the small "mark-up" over substrate costs makes it very important to optimize all parts of the process. This optimization requires a knowledge of basic continuous culture principles as well as a knowledge of microbial physiology, process biochemistry, chemical engineering, and economics. All aspects of the process are interrelated and we feel that no single factor (e.g., yield) should be considered in isolation, but should be related to productivity, capital costs, plant size, cooling costs, etc. For this reason an integrated view should be taken of the whole process (Figure 6).[74]

Undoubtedly, further advances in fermenter design, genetic manipulation of microorganisms, and process control will lead to the development of second generation SCP processes; this development will have a profound influence on continuous-culture applications and biotechnology as well as on the study of microbial physiology. In the review we have touched on some areas of such interest. Examples are the (1) possible evolutionary importance of yield and ways in which the yield can be maximized, (2) the nature of "uncoupling" or "slip" reactions in microbial cultures, and (3) the physiology of microorganisms at interfaces in the fermenter, in the gas head-space liquid droplets, and under physical conditions found in the fermenter. There will undoubtedly be further developments in our understanding of these and other areas. For example, an increase in the utilization of renewable substrates for SCP production and the further development of combined SCP production-effluent treatment units exemplified by the "Pekilo" process[75] for fungal biomass production from sulfite waste liquor. With reference to fermenter design this will advance further. There is the development of the ICI pressure cycle fermenter and we will undoubtedly see further developments which will minimize the power input required for a given rate of oxygen transfer. In all these cases the development must be as an integrated process and the chemical engineer must be aware of how the physical conditions prevailing in the fermenter can influence the physiology of the microorganisms.

V. APPENDIX

In order to update this review some of the recent developments in S.C.P. production will be highlighted. The reader is referred to a number of overviews (1 to 6) that have been published since this article was completed. These reviews also discuss the economics of the various routes. I.C.I. has built a 1500 m^3 pressure cycle fermenter for the continuous production of S.C.P. from methanol (which is produced chemically from methane). Several important problems associated with scale-up have been highlighted.[99] The importance of maintaining good mixing of methanol is emphasized by the presence of two methanol addition points per m^3 of reactor, i.e. a total of 3000 addition points! Clearly the overall economics of S.C.P. production are very sensitive to the yield factor to justify such expensive engineering. We have already discussed how poor mixing of methanol gives a depression in yield. Also the plant includes effluent treatment facilities capable of dealing with the whole fermenter contents,this is as a standby. Normally the effluent treatment problem is one of removing a low concentration of organic carbon from the culture supernatant before it is discharged.

Genetic engineering techniques have been used to increase the yield of bacteria on methanol.[99,100] I.C.I. have succeeded in removing the ATP-dependent GS pathway of ammonia assimilation and replacing it with the ATP independent GDH from *Escherichia coli.* They claim to get the expected (as we have discussed in the main text) 5 to 7% increase in yield on methanol in their constructed organism; whether this genetically engineered strain is stable or whether this increased yield will be realized in the full scale production plant remains to be seen. A much more dramatic increase in yield on methanol, perhaps up to 40%, for a ribulose monophosphate pathway bacterium may be realized if 2 ATP rather than 1 ATP were produced during methanol oxidation.[101] This might be achieved by replacement of the existing pyrroloquinoline quinone linked[102] methanol dehydrogenase by a FAD^+, or even NAD^+ linked dehydrogenase. Earlier suggestions that ribulose monophosphate pathway bacteria could be intrinsically reductant –, and not energy-limited, because of the fixation of large amounts of CO_2 and its reduction to the level of cell material are probably not true, at least in *Pseudomonas* C. Careful experimentation[11] has indicated that a maximum of only 11% of cell carbon is incorporated from CO_2. The fixation and reduction of this amount of CO_2 would not be a serious drain on the reductant supply.

Recent experimental work and modelling of the continuous methane-based natural gas route[104] has emphasized the importance of mixed cultures when the natural gas contains higher alkanes. Even with low concentrations of ethane and propane relative to methane there will be a need to remove the products of higher alkane "co-oxidation" via added heterotrophic bacteria. Such a continuous process with water recycle becomes very complex with respect to the interactions between various parameters,[104] e.g. between gas composition, rate of gas mass-transfer, biomass productivity, liquid recycle ratio and concentration of co-oxidation products.

Finally we would like to briefly mention dinitrogen fixation and S.C.P. production. At first sight it might seem attractive to convert atmospheric dinitrogen directly into protein via a dinitrogen-fixing S.C.P. process with methane or methanol as the energy source. However, surprisingly the energetics are not too favorable. First, when fixing dinitrogen, because of the reductant and energy requirement for dinitrogen fixation, the yields (Table 1) are lower than when assimilating ammonia, hence for a given productivity there is a greater oxygen requirement and heat output. Second, for the given observed growth yields for nitrogen-fixing methane utilizers the energy cost, simply based on the consumption of methane and ignoring process energy costs, for the production of a ton of fixed nitrogen is 718×10^6 B.T.U.'s which can be compared to a value (including process energy costs) of 35×10^6 B.T.U.'s for the production of a ton of fixed nitrogen from methane and dinitrogen via the conventional chemical route.[105] This is a very marked difference in energy utilization even allowing for the fact that one is comparing nitrogen as ammonia with nitrogen as protein.

ACKNOWLEDGMENTS

We would like to thank Dr. G. Hamer for his encouragement while we wrote this review, and all the staff at Shell Research Limited for any discussions.

REFERENCES

1. **Wilson, C.**, cited by Scrimshaw, N. S., in Introduction *Single Cell Protein*, Mateles, R. I. and Tannebaum, S. R., Eds., MIT Press, Cambridge, 1968, 7.
2. **Maclennan, D. G.**, Single cell protein from starch in *Continuous Culture 6: Application and New Fields*, Dean, A. C. R., Ellwood, D. C., Evans, C. G. T., and Melling, J., Eds, Ellis Horwood Ltd., Chichester, England, 1976, 69.
3. **Herbert, D.**, Stoichiometric aspects of microbial growth, in *Continuous Culture 6: Application and New Fields*, Dean, A. C. R., Ellwood, D. C., Evans, C. G. T., and Melling, J., Eds., Ellis Horwood Ltd., Chichester, England, 1976, 1.
4. **Wang, H. Y., Cooney, C. H., and Wang, D. I. C.**, Computer-aided bakers' yeast fermentations, *Biotechnol. Bioeng.*, 14, 69, 1977.
5. **Litchfield, J. H.**, Comparative technical and economic aspects of single cell protein processed, *Adv. Appl. Microbiol.*, 22, 267, 1977.
6. **Dimmling, W. and Seipenbusch, R.**, Raw materials for the production of SCP, *Process Biochem.*, 13, 9, 1978.
7. **Sahm, H., Roggenkamp, R., Wagner, F., and Hinklemann, W.**, Microbodies in methanol grown *Candida boidinii*, *J. Gen. Microbiol.*, 88, 218, 1975.
8. **Senior, P. J. and Dawes, E. A.**, Energy reserve polymers in microorganisms, *Adv. Microb. Physiol.*, 10, 135, 1973.
9. **Waslien, C. I., Calloway, D. H., and Margen, S.**, Human tolerance to bacteria as food, *Nature (London)*, 221, 84, 1969.
10. **Linton, J. D. and Cripps, R. E.**, The occurrence and identification of intracellular polyglucose storage granules in *Methylococcus* NCIB 11083, grown in chemostat culture on methane, *Arch. Microbiol.*, 117, 41, 1978.
11. **Vary, P. S. and Johnson, M. J.**, Cell yields of bacteria grown on methane, *Appl. Microbiol.*, 15, 1473, 1967.
12. **Cremieux, A., Chevalier, J., Combet, M., Dumenil, G., Parlouar, D., and Ballerini, D.**, Mixed culture of bacteria utilizing methanol for growth, *Eur. J. Appl. Microbiol.*, 4, 1, 1977.
13. **Whittenbury, R., Phillips, K. C., and Wilkinson, J. F.**, Enrichment, isolation and some properties of methane-utilizing bacteria, *J. Gen. Microbiol.*, 61, 205, 1970.
14. **Johnson, M J.**, Techniques for selection and evaluation of cultures for biomass production, in *Fermentation Technology Today*, Proc. IV IFS, Terui, G., Ed., Society of Fermentation Technology Japan, 1972, 473.
15. **Snedcor, B. and Cooney, C. L.**, Thermophilic mixed culture of bacteria utilizing methanol for growth, *Appl. Microbiol.*, 27, 1112, 1974.
16. **Ballerini, D., Parlouar, D., Lapeyronnie, M., and Sri, K.**, Mixed culture of bacteria utilizing methanol for growth. 1. Isolation and identification, *Eur. J. Appl. Microbiol.*, 4, 11, 1977.
17. **Wilkinson, T. G., Topiwala, H. H., and Hamer, G.**, Interactions in a mixed bacterial population growing on methane in continuous culture, *Biotechnol. Bioeng.*, 16, 41, 1974.
18. **Jannasch, H. W.**, Enrichment of aquatic bacteria in continuous culture, *Arch. Mikrobiol.*, 59, 165, 1967.
19. **Novick, A. and Szilard, L.**, Description of the chemostate, *Science*, 112, 715, 1950(a).

20. **Novick, A. and Szilard, L.**, Experiments with the chemostat on spontaneous mutation of bacteria, *Proc. Natl. Acad. of Sci. U.S.A.*, 36, 708, 1950(b).
21. **Sheehan, B. T. and Johnson, M. J.**, Production of bacterial cells from methane, *Appl. Microbiol.*, 21, 511, 1971.
22. **Harder, W. and Kuenen, J. G.**, A review. Microbial selection in continuous culture, *J. Appl. Bacteriol.*, 43, 1, 1977.
23. **Kuenen, J. G. and Veldkamp, H.**, De Chemostaat als Lulpmiddel bij de Stodie van de Oecologie van Bacterien, *Vakbl. Biol.*, 6, 100, 1973.
24. **Veldkamp, H. and Kuenen, J. G.**, The chemostat as a model system for ecological studies, *Bull. Ecol. Res. Comm. (Stockholm)*, 17, 347, 1973.
25. **Harrison, D. E. F.**, Making protein from methane, *Chem. Technol.*, 6, 570, 1976.
26. **Harrison, D. E. F., Wilkinson, T. G., Wren, S. J., and Harwood, J. H.**, Mixed bacterial cultures as a basis for continuous production of single cell protein from C-1 compounds in continuous culture, in *Continuous Culture 6: Applications and New Fields,* Dean, A. C. R., Ellwood, D. C., Evans, C. G. T., and Melling, J., Eds., Ellis Horwood Ltd., Chichester, England, 1975, 122.
27. **Goldberg, I.**, Production of SCP from methanol-yield factors, *Process Biochem.*, 12, 12, 1977.
28. **Dostalek, M. and Molin, N.**, Studies of biomass production of methanol oxidising bacteria, in *Single Cell Protein,* Vol. 2, Tannebaum, S. R., and Wang, D. I. C., Eds., MIT Press, Cambridge, 1975, 395.
29. **Linton, J. D. and Buckee, J. C.**, Interactions in a methane utilising mixed bacterial culture in a chemostat, *J. Gen. Microbiol.*, 101, 219, 1977.
30. **Harrison, D. E. F., and Wren, S. J.**, Mixed microbial cultures as a basis for future fermentation processes, *Process Biochem.*, 30, 11, 1976.
31. **Drozd, J. W., Khosrovi, B., Downs, J., Bailey, M. L., and Barnes, L. J.**, Biomass production from natural gas, in 7th Int. Symp. on Continuous Cultivation of Micro-organisms, Sikyta, B., Ed., Prague, 1978.
32. **Knecht, R., Präve, P., Seinpenbusch, M., and Sukatsch, D. A.**, Microbiology and biotechnology of SCP produced from n-paraffin, *Process Biochem.*, 12, 11, 1977.
33. **Shiloach, J. and Bauer, S.**, High-yield growth of *E. coli* at different temperatures in a bench scale fermenter, *Biotechnol. Bioeng.*, 17, 227, 1975.
34. **VanDijken, J. P. and Harder, W.**, Growth yields of micro-organisms on methanol and methane. A theoretical study, *Biotechnol. Bioeng.*, 17, 15, 1975.
35. **Downs, J., Drozd, J. W., Khosrovì, B., Linton, J. D., and Barnes, L. J.**, An analysis of growth energetics in *Methylococcus* NCIB 11083, *Proc. Soc. Gen. Microbiol.*, 5, 45, 1978.
36. **Herbert, D.**, The chemical composition of micro-organisms as a function of their environment. In microbial reaction to environment, *Symp. Soc. Gen. Microbiol.*, 11, 391, 1961.
37. **Hedenskog, G.**, Properties and composition of single cell protein, influence of processing, in *Biochemical Aspects of New Protein Food,* Vol. 44, 11th FEBS Meeting, Adler-Nissen, J., Eggum, B. O., Munck, L., and Olsen, H. S., Eds., Pergamon Press, Oxford, 1978, 73.
38. **Pirt, S. J.**, The maintenance energy of bacteria in growing cultures, *Proc. R. Soc. London, Ser. B,* 163, 224, 1965.
39. **Abbott, B. J. and Clamen, A.**, The relationship of substrate, growth rate, and maintenance coefficient to single cell protein production, *Biotechnol. Bioeng.*, 15, 117, 1973.
40. **Harrison, D. E. F. and Loveless, J. E.**, The effect of growth conditions on respiratory activity and growth efficiency in facultative anaerobes grown in chemostat culture, *J. Gen. Microbiol.*, 68, 35, 1971.
41. **Humphrey, A. E.**, Product outlook and technical feasibility of SCP, in *Single Cell Protein,* Vol. 2, Tannebaum, S. R. and Wang, D. I. C., Eds., MIT Press, Cambridge, 1975, 1.
42. **Harrison, D. E. F., Topiwala, H. H., and Hamer, G.**, Yield and productivity in single cell protein production from methane and methanol, in *Fermentation Technology Today,* Proc. IV IFS, Terui, G., Ed., Society of Fermentation Technology Japan, 1972, 491.
43. **Harrison, D. E. F.**, Physiological effects of dissolved oxygen tension and redox potential of growing populations of micro-organisms, *J. Appl. Chem. Biotechnol.*, 22, 417, 1972.
44. **Neijssel, O. M. and Tempest, D. W.**, Bioenergetic aspects of aerobic growth of *Klebsiella aerogenes* NCTC 418 in carbon-limited and carbon sufficient chemostat cultures, *Arch. Microbiol.*, 107, 215, 1976.
45. **Stouthamer, A. H.**, Energetic aspects of the growth of micro-organisms, in *Microbial Energetics,* 27th Symp. Soc. for General Microbiology. Haddock, B. A. and Hamilton, W. A., Eds., Cambridge University Press, London, 1977, 285.
46. **Brooks, J. D. and Meers, J. L.**, The effect of discontinuous methanol addition on the growth of a carbon-limited culture of *Pseudomonas, J. Gen. Microbiol.*, 77, 513, 1973.

47. **Topiwala, H. H. and Khosrovi, B.**, Water recycle in biomass production processes, *Biotechnol. Bioeng.*, 20, 73, 1978.
48. **Reuss, M., Sahm, H., and Wagner, F.**, Mikrobielle Protein-Gewinnung auf Methanol-Basis, *Chem. Ing. Tech.*, 46, 669, 1974.
49. **Monod, J.**, La technique de culture continue: théorie et applications, *Ann. Inst. Pasteur Paris*, 79, 390, 1950.
50. **Herbert, D., Elsworth, B., and Telling, R. C.**, The continuous culture of bacteria; a theoretical and experimental study, *J. Gen. Microbiol.*, 14, 601.
51. **Harrison, D. E. F.**, Studies on the affinity of methanol and methane-utilising bacteria for their carbon substrates, *J. Appl. Bacteriol.*, 36, 301, 1973.
52. **Pirt, S. J.**, *Principles of Microbe and Cell Cultivation*, Blackwells Scientific Publications, Oxford, 1975.
53. **Romanovskaya, V. A., Malashenko, Yu, R., Sokolov, I. G., and Kryshtab, T. P.**, The competitive inhibition of the microbial oxidation of methane by ethane, in *Microbial Production and Utilisation of Gases*, Schlegel, H. G., Gottschalk, G., and Pfenning, N., Eds., E. Goltze, K. G. Göttingen, West Germany, 1976, 345.
54. **Drozd, J. W., Bailey, M. L., and Godley, A.**, The physiology of mixed bacterial cultures grown on natural gas, *Proc. Soc. Gen. Microbiol.*, 4, 26, 1976.
55. **Linton, J. D., Bull, A. T., and Harrison, D. E. F.**, Determination of the apparent Km for oxygen of *Beneckea natriegens.* Using the respirograph technique, *Arch. Microbiol.*, 114, 111, 1977.
56. **Cooney, C. L., Wang, D. I. C., and Mateles, R. I.**, Measurement of the heat evolution and correlation with oxygen consumption during microbial growth, *Biotechnol. Bioeng.*, 11, 269, 1969.
57. **Weast, R. C., Ed.**, *Handbook of Chemistry and Physics*, 51st ed., CRC Press, Cleveland, 1970, D-217.
58. **Cooney, C. L.**, Engineering consideration in the production of single-cell protein from methanol, in *Microbial Growth on C-1 Compounds*, The Society for Fermentation Technology, Osaka, Suita, Japan, 1975, 183.
59. **Anthony, C.**, The prediction of growth yields in methylotrophs, *J. Gen. Micobiol.*, 104, 91, 1978.
60. **Taylor, S.**, Evidence for the presence of ribulose 1,5-biphosphate carboxylase and phosphoribonuclease in *Methylococcus capsulatus*, (Bath), *FEMS Microbiol. Lett.*, 2, 305, 1977.
61. **Linton, J. D. and Stephenson, R. J.**, A preliminary study on growth yields in relation to the carbon and energy content of various organic growth substrates, *FEMS Microbiol. Lett.*, 3, 95, 1978.
62. **Tonge, C. M., Harrison, D. E. F., and Higgins, I. J.**, Purification and properties of the methane mono-oxygenase enzyme system from *Methylosinus trichosporium* OB3b, *Biochem. J.*, 161, 333, 1977.
63. **Colby, J. and Dalton, H.**, Some properties of a soluble methane mono-oxygenase from *Methylococcus capsulatus* strain, Bath, *Biochem. J.*, 157, 495, 1976.
64. **Linton, J. D. and Vokes, J.**, Growth of the methane utilising bacterium *Methylococcus* NCIB 11083 in mineral salts medium with methanol as the sole source of carbon, *FEMS Microbiol. Lett.*, 4, 125, 1978.
65. **Harrison, D. E. F. and Topiwala, H. H.**, Transient and oscillatory states of continuous culture, *Adv. Biochem. Eng.*, 3, 167, 1974.
66. **Drozd, J. W., Godley, A., and Bailey, M. L.**, Ammonia oxidation by methane oxidising bacteria, *Proc. Soc. Gen. Microbiol.*, 5, 66, 1978.
67. **Tempest, D. W.**, Quantitative relationships between inorganic cations and anionic polymers in growing bacteria, in *Microbial Growth*, 19th Symp. of the Society for General Microbiology. Meadow, P. M. and Pirt, S. J., Eds., Cambridge University Press, London, 1969, 87.
68. **Light, P. A.**, Influence of environment on mitochondrial function in yeast, *J. Appl. Chem. Biotechnol.*, 22, 509, 1972.
69. **Mateles, R. I. and Battat, E.**, Continuous culture used for media optimization, *Appl. Microbiol.*, 28, 901, 1974.
70. **Gow, J. S., Littlehailes, J. D., Smith, S. R. L., and Walter, R. B.**, SCP production from methanol: bacteria, in *Single Cell Protein*, Vol. 2, Tannebaum, S. R. and Wang, D. I. C., Eds., MIT Press, Cambridge, 1975, 370.
71. **Faust, U., Präve, P., and Sukatsch, D. A.**, Continuous biomass production from methanol by *Methylomonas clara*, *J. Ferment. Technol. (Japan)*, 55, 609, 1977.
72. **Hamer, G.**, Wall growth and its significance in large fermenters used for the cultivation of single celled micro-organisms, *Biotechnol. Bioeng., Symp.*, 4, 565, 1973.
73. **Hamer, G.**, Discussion on entrained droplets in fermenters used for the cultivation of single celled micro-organisms, *Biotechnol. Bioeng.*, 14, 1, 1972.
74. **Topiwala, H. H.**, The relevance of the research on single cell protein manufacturing processes, in *Microbial Growth on C_1-Compounds*, The Society for Fermentation Technology, Osaka, Japan, 1975, 199.

75. **Romantschuk, H. and Lehtomaki, M.,** Operational experiences of first full scale Pekilo SCP-Mill application, *Process Biochem.*, 13 16, 1978.
76. **Goldberg, I., Rock, J. S., Ben-Bassat, A., and Mateles, R. I.,** Bacterial yields on methanol, methylamine, formaldehyde and formate, *Biotechnol. Bioeng.*, 18, 1657, 1976.
77. **Dostalek, M., Haggstrom, L., and Molin, N.,** Optimisation of biomass production from methanol, in *Fermentation Technology Today*, Proc. IV IFS, Terui, G., Ed., Society of Fermentation Technology, Japan, 1972, 497.
78. **Minami, K., Yamamura, M., Shimizu, S., Ogawa, K., and Sekine N.,** A new methanol-assimilating high productive, thermophilic yeast, *J. Ferment. Technol. (Japan)*, 56, 1, 1978.
79. **Levine, C. W. and Cooney, C. L.,** Isolation and characterisation of a thermotolerant methanol-utilising yeast, *Appl. Microbiol.*, 26, 982, 1973.
80. **Bewersdorff, M. and Dostalek, M.,** The use of methane for production of bacterial protein, *Biotechnol. Bioeng.*, 13, 49, 1971.
81. **Harwood, J. H. and Pirt, S. J.,** Quantitative aspects of growth of the methane oxidising bacterium *Methylococcus capsulatus* on methane in shake flasks and continuous chemostat culture, *J. Appl. Bacteriol.*, 35, 597, 1972.
82. **Grigorian, A. N. and Lalov, V. V.,** Studies in cultivation of micro-organisms on natural gas, in Natural Gas Processing and Utilisation Conference, Dublin, 1977, *Inst. Chem. Eng. Symp. Ser.*, 44, 1976.
83. **Barnes, L. J., Drozd, J. W., Harrison, D. E. F., and Hamer, G.,** Process considerations and techniques specific to protein production from natural gas, in *Microbial Production and Utilisation of Gases*, Schlegel, H. G., Gottschalk, G., and Pfennig, N., Eds., E. Goltze, K. G. Göttingen, West Germany, 1976, 301.
84. **Hamer, G., Harrison, D. E. F., Harwood, J. H., and Topiwala, H. H.,** *SCP Production from Methane*, Proc. 2nd Int. Conf. on SCP, MIT, Tannebaum, S. R. and Wang, D. I. C. Eds., MIT Press, Cambridge, 1973, 357.
85. **Evashikov, E. I., Romanovskaya, V. A., and Malashenko, R.,** The Inhibitors of Growth of Obligate Methylothophs, Proc. 2nd Int. Cont. on SCP, MIT, Tannebaum, S. R. and Wang, D. I. C., Eds., MIT Press, Cambridge, 1973, 419.
86. **Leadbetter, E. F. and Foster, J. W.,** Bacterial oxidation of gaseous alkanes, *Arch. Microbiol.*, 35, 92, 1960.
87. **Quayle, J. R.,** The metabolism of one-carbon compounds by microorganisms, *Adv. Microb. Physiol.*, 7, 119, 1972.
88. **Birckenstaedt, J. W., Faust, U., and Sambeth, W.,** Production of SCP from n-paraffin-processes and products, *Process Biochem.*, 7, 12, 1977.
89. **Calvert, R. and Hamer, G.,** unpublished observations.
90. **Wren, S. J. and Harrison, D. E. F.,** The role of heterotrophic bacteria in a mixed culture growing on methanol, *Proc. Soc. Gen. Microbiol.*, 4, 29, 1976.
91. **Drozd, J. W., Linton, J. D., Downs, J., and Stephenson, R. J.,** An in situ assesssment of the specific lysis rate in continuous cultures of *Methylococcus* sp. (NC1B 11083) grown on methane, *FEMS Microbiol. Lett.*, 4, 311, 1978.
92. **Bailey, M. L., Downs, J., and Drozd, J. W.** Nitrogen metabolism in *Methylococcus* NC1B 11083, *Proc. Soc. Gen. Microbiol.*, 5, 65, 1978.
93. **Hamer, G. and Hamdan, I. Y.,** Protein Production by Micro-organisms, *Chem. Soc. Rev.* 8, 143—170, 1979.
94. **Hamer, G.,** Technical aspects of single cell protein production from natural gas (methane), in *Proc. of the Regional Seminar on Microbial Conversion Systems for Food and Fodder Production and Waste Management*, Overmire, T. G., Ed., Kuwait Institute for Scientific Research, 1977, 109.
95. **Hamer, G. and Harrison, D. E. F.,** *Single-Cell Protein: The Technology, Economics and Future Potential in Hydrocarbons in Biotechnology*, Harrison, D. E. F., Higgins, I. J. and Watkinson, R., Eds., Institute of Petroleum, London 1980, 59.
96. **Forage, A. J. and Righelato, R. C.,** Biomass from carbohydrates, in *'Economic Microbiology', Microbial Biomass*, Vol. 4, Rose, A. H., Ed., Academic Press, London, 1979, 289.
97. **Hamer, G.,** Biomass from natural gas, in *'Economic Microbiology', Microbial Biomass*, Vol. 4, Rose, A. H., Ed., Academic Press, London, 1979, 315.
98. **Levi, J. D., Shennan, J. L., and Ebbon, G. P.,** Biomass from liquid n-alkanes, in *'Economic Microbiology', Microbial Biomass*, Vol. 4, Rose, A. H., Ed., Academic Press, London, 1979, 361.
99. **Senior, P. J. and Windass, J.,** The I.C.I. single cell protein process, *Biotechnol. Lett.*, 2, 205, 1980.
100. **Windass, J. D., Worsey, M. J., Pioli, E. M., Pioli, D., Barth, P. T., Atherton, K. T., Dart, E. C., Byrom, D., Powell, K., and Senior, P. J.,** Improved conversion of methanol to single cell protein by *Methylophilus methylotrophus*, *Nature*, 287, 396, 1980.

101. **Drozd, J. W. and Wren, S. J.,** Growth energetics in the production of bacterial single cell protein from methanol, *Biotechnol. Bioeng.,* 22, 353, 1980.
102. **Duine, J. A., Frank, J., and Verwiel, P. E. J.,** Structure and activity of the prosthetic group of methanol dehydrogenase, *Eur. J. Biochem.,* 108, 187, 1980.
102. **Ben-Bassat, A., Goldberg, I. and Mateles, R. I.,** *J. Gen. Microbiol.,* 116, 213, 1980.
104. **Drozd, J. W. and McCarthy, P. W.,** Mathematical model of hydrocarbon oxidation, in *3rd Int. Symp. on Microbial Growth on C_1- Compounds; Sheffield 1980,* Dalton, H. D., Ed., Heyden and Son, London, in press.
105. **Nichols, D. and Blouin, G. M.,** Ammonia fertiliser from coal, *Chemtech,* 9, 512, 1979.

Chapter 7

TRANSIENT PHENOMENA IN CONTINUOUS CULTURE*

Charles L. Cooney, H. Michael Koplov, and Margareta Häggström

TABLE OF CONTENTS

I. Introduction 144

II. Transients in Response to Environmental Changes 144
A. Introduction 144
B. Nutritional Changes 145
1. Dilution Rate Changes 145
2. Pulse Addition of Limiting Nutrients 146
3. Step Changes in the Limiting Nutrient Concentration 149
C. Environmental Changes 151
1. Temperature Change 151
2. Inhibitor Addition 152
3. Carbon Catabolite Repression 153
4. Aerobic/Anaerobic Transitions 153
a. Metabolic Pool Size 154
b. Product Formation 156
D. Summary Discussion on Environmental Perturbation 156

III. Enzyme Synthesis during Transient Responses 157
A. Introduction 157
B. Molecular Events Following Perturbation 157
C. Enzyme Synthesis as a Means for Response 159

IV. Understanding and Applications of Transient Phenomena 163
A. Introduction 163
B. Understanding Transient Phenomena 163
C. Application of Transient Phenomena 164

References 165

* This chapter was submitted in April 1979.

I. INTRODUCTION

The technique of continuous culture has become an important tool not only for studies of microbial physiology, but also for use in the production of microbial products such as single cell protein, enzymes, and organic acids.[1] By far, the major focus in the use of continuous culture has been in the steady state. The reason for this is that steady-state operation allows the investigator to take advantage of the time invariant properties of steady state, and the ability to manipulate single independent variables during steady-state growth. Furthermore, when continuous culture is used for production purposes, it is desired to maintain the cells under optimal environmental conditions which, in this case, characterize an optimum steady state. It is interesting, however, to consider the nonsteady state or transient behavior of microorganisms in continuous culture. After all, a steady state can only be obtained after a transitionary period. Interesting transient phenomena in continuous culture, however, go far beyond the understanding of the transients between steady states. The use of transients is becoming an important tool in studying metabolic events.[2-4] An understanding of transient phenomena is important to the development and application of dynamic control in continuous culture processes.[5-7] Lastly, the inducement of transient behavior can provide a useful way of studying events that occur in nature.[8] That is to say that the "steady state" in natural systems represents the average of many transient phenomena occurring over short periods of time and in local ecologic niches.

The focus in this chapter is to examine some of the work that has been done on transient systems in continuous culture. Initial attention will be paid to a description of the response of microbial systems to changes in the environment. After this, close attention will be paid to enzyme synthesis during transients, in an attempt to understand transient phenomena on a microscale. Lastly, consideration will be given to the practical aspects of transient behavior along with identification of some of the unknowns in our understanding of transient phenomena in continuous culture.

For the purpose of this discussion, transient phenomena are considered to be events that follow either premeditated or accidental alterations in the growth environment. Such alterations may involve chemical changes, physical changes, or organism changes in the case of mixed cultures. Steady state is a condition in which selected objective functions, e.g., cell mass, limiting or non-limiting nutrient concentration, enzyme level, product concentration, etc. are invariant with time. Thus, steady state is operationally defined. Transient state presumes that one or more of these objective functions changes with time.

In batch fermentations, specific growth and product formation rates can be maintained at constant and this is called "balanced growth." It is not, however, steady state, and substrates, cell, and product concentrations will vary with time. This example with batch culture illustrates the need to be careful in defining steady state and transient growth. Here, the discussion is restricted to continuous culture and an analysis of transients induced therein. The reader is also referred to the reviews by Harrison and Topiwala[9] and Koplove and Cooney[10] for additional discussions of transient phenomena.

II. TRANSIENTS IN RESPONSE TO ENVIRONMENTAL CHANGE

A. Introduction

Attention here is directed to transient behavior in continuous culture and its effects on cell mass and product formation. The chemostat has been used extensively for studying transient state behavior of cultures at both the population and metabolic level. A study of transient response of microorganisms is a useful source of information on mechanisms regulating growth and metabolism. Transient studies are also important in understanding the dynamic behavior of continuous cultures which is necessary for scale-up and operation of fermentation processes. Transient and oscillatory phenomena in continuous culture can be initiated by interactions between component species in mixed populations of organisms in an otherwise constant environment, or by changes in the environment. The focus here is primarily on transient response to environmental changes.

B. Nutritional Changes

Nutritional changes that result in transient behavior of a culture can involve: (1) step changes in dilution rate, (i.e., medium addition rate at constant volume), (2) pulse addition of one or more nutrients, (3) change in feed medium concentration or composition. According to Monod,[11,12] changes in substrate concentration(s) will affect the specific growth rate (μ) according to the relationship

$$\mu = \mu_{max} \cdot \frac{S}{K_s + S} \quad (1)$$

where μ_{max} is the maximum specific growth rate (hr^{-1}), K_s the half rate constant (mg/ℓ), and S the limiting substrate concentration (mg/mℓ).

Thus, according to this model, growth is regulated by the limiting substrate, and the quantitative effect of nutritional changes will be determined by the ratio of S/K_s.

1. Dilution Rate Changes

Mateles et al.[13] and Ryu[14] examined the transient response of *Escherichia coli* after shifts in dilution rate in a nitrogen-limited chemostat. They found that the magnitude and the direction of the shifts were important factors in determining the response of the culture. A large increase in dilution rate (> 0.2 hr^{-1}) resulted in a transient period of several hours during which the limiting substrate concentration increased before the new growth rate was reached. Small changes in dilution rate (< 0.2 hr^{-1}) resulted in an immediate adjustment by the cells to the new growth rate. These results suggested that the organisms had a reserve biosynthetic capacity that would allow for a small (< 0.2 hr^{-1}) immediate increase in growth rate. It is important to note from their work that the Monod model does not describe growth as a function of nutrient concentration under transient conditions. This conclusion was also confirmed by Cooney and Wang[15] in studies with *Enterobacter aerogenes* grown under nitrogen and nitrogen-phosphate limitation. The Monod model predicts a faster response to environmental changes than actually occurs; this point, to be discussed later, suggests that the mechanism of growth limitation is not the same under both steady state and transient conditions. In this context, changes occurring in cell macromolecular composition, enzyme level, and cell yield will be examined during transient response.

For instance, Koga and Humphrey[16] have shown that variation in cell yield on the limiting nutrient can lead to damped oscillations before a new steady state is reached after a disturbance. Studying *Saccharomyces cerevisiae* in a nitrogen-limited chemostat, Gilley and Bungay[17] observed damped oscillations after a shift up in dilution rate from 0.125 to 0.270 hr^{-1}; small changes in the yield coefficient could be observed. They reported that the time required for the oscillations to damp out was a function of the magnitude of the dilution rate shift, dilution rate range, and limiting nutrient concentration.

Mor and Fiechter[18] using a carbon-limited chemostat of *S. cerevisiae* followed several parameters during the transient period following a shift-up in dilution rate. They observed damped oscillations in cell and substrate concentration, oxygen uptake rate and carbon dioxide release at low dilution rates (< 0.07 hr^{-1}). At higher dilution rates (0.10 to 0.17 hr^{-1}), however, oscillations were seen in the gas exchange rates, but not in other parameters. The lower the dilution rate, the lower was the rate of damping. The difference in results from those obtained by Gilley and Bungay[17] may be explained by the fact that the relative magnitude of the shift used was higher at the low dilution rates than at the higher dilution rates.

The behavior of cells during transients following shifts in dilution rate has been further characterized. Regan and Roper[19] showed for *S. cerevisiae* that a step increase in dilution rate induced budding and partial cell synchrony. Meers[20] studied the response of a mixed culture of *Bacillus subtilis* and *Torula utilis;* a shift up in dilution rate in a magnesium-limited chemostat resulted in replacement of *T. utilis* by *B. subtilis,* but the yeast outgrew the bacterium at lower growth rates. Lee et al.[21] observed similar changes in mixed cultures of *Lactobacillus plantarum* and *Propionibacterium shermanii* after shifts in dilution rate. A mixed culture of *Acetobacter suboxydans* and *S. carlsbergensis* showed oscillations in both substrate and cell concentrations after a step increase in dilution rate. In this study, the limiting carbon source, mannitol, was oxidized by the bacterium to fructose, which was consumed by the yeast. No fluctuations were observed when the same step increase was performed in a pure culture of *A. suboxydans,* but, in the mixed culture, the oscillations in fructose and yeast concentration were the most pronounced. These results indicate that the instability of the mixed-culture may be caused by a feedback mechanism from the yeast to the bacterium. Cunningham and Maas[22] reported damped oscillations in cell number following a rapid change in dilution rate in a nitrogen-limited chemostat of the green flagellate *Clamydomonas reinhardii.* During the transient period, the specific growth rate and the average amount of limiting nutrient per cell oscillate with about the same period, but with a phase lag. They hypothesized a partial uncoupling of the system controlling cell division from changes in environmental limiting nutrient concentration. Cell division appeared to be controlled by the amount of intracellular nitrogen. Storage of nitrogen may occur at high growth rates and there is a time lag between fluctuations in intracellular nitrogen content and in growth rate.

2. Pulse Additions of Limiting Nutrients

A second method to induce transients in continuous culture is through the pulse addition of the limiting nutrient to temporarily remove the nutrient limitation. Assuming that all other nutrients are in excess, this type of stress will permit the cell to respond at its maximum possible rate, dictated not initially by a nutrient concentration, but by some intracellular constraint, e.g., the level of some key enzyme(s), the availability of energy, or reducing equivalents for biosynthesis.

Welles and Blanch[23] conducted pulse feeding experiments in an anaerobic glucose-limited culture of *S. cerevisiae.* They showed that by continuous pulse feeding at a frequency of 0.25 (c/min) they could increase the ethanol yield from glucose by 50% above the value obtained in continuous steady-state carbon-limited culture. These results may be explained by the budding process that occurs rapidly at high growth rates and results in higher energy requirement; this means a greater flux of ethanol production and will result in an increased product yield. The results may also be explained according to a hypothesis by Brookes and Meers[6] that more energy may be required for the continual synthesis and degradation of enzymes that are necessary for the survival of organisms in a changing environment. Similar studies have been carried out by Vairo et al.[24, 25] to simulate transient responses observed in domestic sewage plants. Periodic variations in feeding mash concentration or dilution rate (1 c/24 hr) resulted in cyclic variations in cell yield and specific growth rate. However, the average yield coefficient for cell mass formation was essentially equal to the yield coefficient obtained during steady-state growth. The average specific growth rate during the transients equaled the dilution rate.

Leung and co-workers[4, 26] have studied the transient growth behavior of a glucose-limited chemostat of *Streptococcus mutans* after a pulse of glucose. The transient response shown in Figure 1 is characterized by a rapid increase in production of lactic acid. The maximum specific productivity observed during the transient was dependent on the growth rate prior to the glucose pulse, with faster growing cultures exhibiting a greater increase in specific productivity than slow growing cultures.[4] After the pulse, the yield coefficient for lactic acid (gram of lactic acid per gram glucose) increased from a steady state value averaging 0.4 to around 0.9. When inorganic phosphate was added with the glucose pulse, the transient lactic acid production was reduced in comparison with glucose added alone.[27] Lactate dehydrogenase (LDH), a key enzyme in the pathway from glucose to lactate, is activated by fructose-1,6 diphosphate (FDP), and this activation has been shown to be inhibited by inorganic phosphate. Addition of the protein synthesis inhibitor chloramphenicol with the glucose pulse also resulted in a lower increase in lactic acid production.[27] The phosphate and chloramphenicol effects were additive, suggesting that the transient response of the organism has at least two different points of control. The increase in lactic acid production during the transient is most likely dependent on activation of LDH already present in the cell, as well as on *de novo* synthesis of lactate dehydrogenase and/or other key enzymes.

In the preceding, it was shown that pulse feeding of nutrients may offer an exciting possibility for increasing product yields, e.g., as ethanol and lactic acid production from glucose. However, there are also disadvantages associated with pulse or discontinuous feeding of nutrients. The increase in product yields observed in the anaerobic fermentations described above also results in a decrease in cell yield. In studies with a methanol-limited chemostat of *Pseudomonas methylotropha,* Brooks and Meers[6] observed lower cell yields when discontinuous medium feeding was used, when compared with continuous medium addition. As shown in Figure 2, when the time between nutrient additions was 100s the cell yield fell to about 0.35 g cell mass per gram methanol from an average yield of 0.44 when using continuous nutrient feed; at the same time, the yield of carbon dioxide increased. The pool amino acids, glutamine, and glutamate also varied during the feeding cycle. The authors discuss two possible explanations for the observed decrease in cell yield. First, that the organism, which had been shown to give higher yields when grown under carbon-limited conditions than when carbon is present in excess may inefficiently catabolize excess methanol. Second, that the continuous readjustment of levels of enzyme and key intermediates in a response to the changing environment requires increased energy and decreases the energy available for cell synthesis. Luscombe[28] however, did not observe a decrease in cell yield in a continuous culture of *Arthrobacter globiformis* fed medium in a discontinuous manner. The results, however, may not be contradictory since Luscombe[28] did pulse experiments with long intervals (120s) between the medium additions and at a very low dilution rate (.01 hr^{-1}), where the yield coefficient is already low. The possibility that an effect on cell yield would be observed using longer intervals between the additions at higher dilution rates therefore cannot be ruled out.

Pulse additions have also been used by Cooney and Wang[15] in studying the transient response of *Enterobacter aerogenes* to additions of ammonia to a culture with dual nutrient (nitrogen and phosphate) limitations. Results of such transients are seen in Figure 3. When ammonia was the only growth-limiting nutrient, the organisms were not capable of responding to a pulse addition of ammonia with their theoretical maximum rate as defined by the Monod model. The response pattern showed that the cells exhibited an initial rapid response, and may have a reserve biosynthetic capacity that

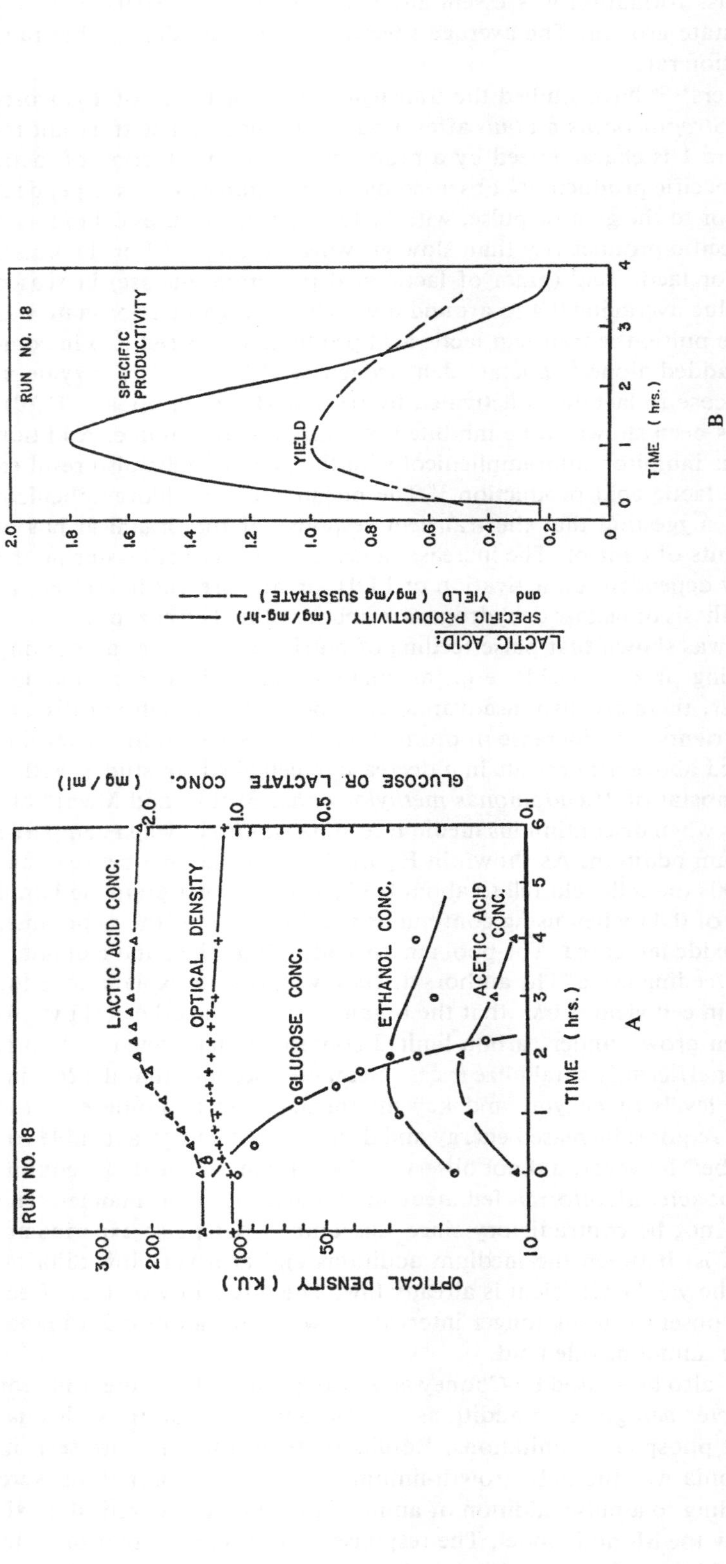

FIGURE 1. The transient response of *Streptococcus mutans* to a pulse of glucose added to a glucose-limited chemostat. (A) shows the time course of events following the pulse and (B) shows the specific productivity and yield of lactic acid production during the transient. (From Cooney, C. L., Leung, J., and Sinskey, A. J., *Microbial Abs.* 799, 1977. With permission.)

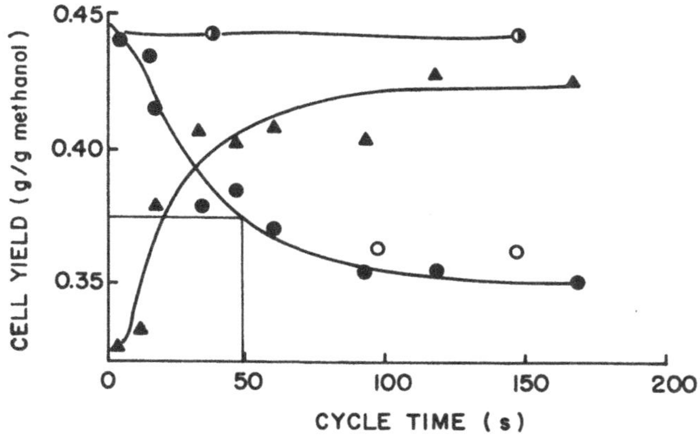

FIGURE 2. The effect of pulsating methanol and nutrient salts fed to a methanol-limited chemostat on the yield of cells (•) and carbon dioxide (▲). When only nutrient salts were added discontinuously, the yield of cells (•) was not altered. Discontinuous addition of methanol alone did affect the cell yield (O). (From Brooks, J. D. and Meers, J. L., *J. Gen. Microbiol.*, 77, 513, 1971. With permission.)

permits immediate adjustment to small changes. However, after a short period of time, there was a lag and then an accelerating response as seen by the uptake of nitrogen. Similar responses have been observed in other transient studies involving step changes in dilution rate.[13,14] An increasing restriction in phosphate availability decreased the ability of the cell to respond to the ammonia pulse. With a dual nitrogen and phosphate limitation, the initial response observed after the pulse addition of ammonia was absent and as the limiting phosphate concentration was decreased, the ability of the cells to respond to ammonia pulses was decreased. The biosynthetic capacity that had been seen under conditions of phosphate excess were felt to be decreased by the phosphate limitation.

Koch and co-workers[2,3] have demonstrated that *E. coli* has a reserve biosynthetic capacity in nitrogen-, phosphate-, and carbon-limited chemostats. After a nutritional shift up, the organism showed an immediate initial response in protein synthesizing capacity, after which they slowly adjusted to the final required level. A sulfate-limited culture did not show this reserve biosynthetic capacity. They postulated a connection between the ability to respond to a change and the RNA synthesizing capacity; this will be discussed in more detail later.

3. Step Changes in the Limiting Nutrient Concentration

Transient behavior, similar to that observed after dilution rate shifts, occurs when the feed medium composition is changed. A step increase in the limiting substrate concentration in the feed medium causes a transient increase in specific growth rate, after which it returns to the original specific growth rate, but at a new and higher cell density.

Regan and Roper[19] investigated the behavior of *Saccharomyces cerevisiae* in a glucose-limited chemostat after changes in glucose concentration in the feed medium. The cell concentration (dry weight) reached its new steady state through a series of steps approximately one generation apart. This observation supported the authors' hypothesis that the step induced a certain degree of synchrony by encouraging the onset of budding. Budding has been shown to be energy intensive and accompanied by rapid utilization of carbohydrate. After the step in glucose supply rate, a large increase in energy production and a small increase in specific glucose consumption rate was observed; this also supported their hypothesis.

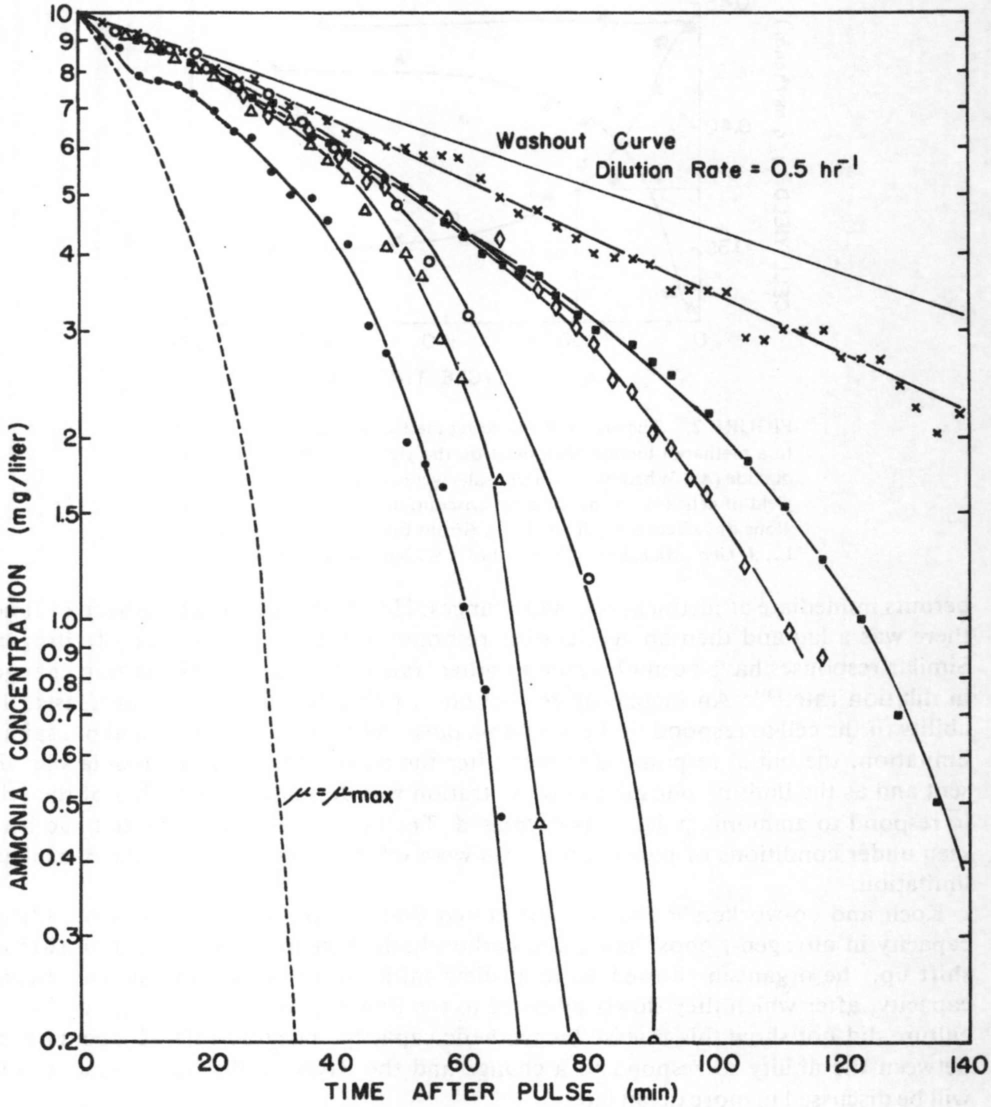

FIGURE 3. Ammonia pulse to a nitrogen-limited chemostat. The curve indicated by $\mu = \mu_{max}$ represents the decrease in ammonia concentration when the cell responds by instantaneously growing at its maximum growth rate. The washout curve indicated by D = 0.5 hr^{-1} represents the rate at which ammonia would be removed from the fermentor if it were not consumed by the cell. The feed ammonia and glucose concentrations were 40 and 1000 mg/ℓ respectively. (From Cooney, C. L. and Wang, D. I. C., *Biotechnol. Bioeng.*, 18, 189, 1976. With permission.)

Curve	Symbol	Feed phosphate (mg/ℓ)	Initial effluent phophate (mg/ℓ)	Final effluent phosphate (mg/ℓ)
1	•	5460	>1000	>1000
2	△	42	10.8	9.3
3	○	30.6	1.2	1.2
4	□	19.8	0.57	—
5	■	19.8	2.1	1.2
6	×	17.5	0.48	0.45

Young and Bungay[5] also studied transients following a step change in feed glucose concentration in a chemostat culture of *S. cerevisiae.* They demonstrated a close relationship between cellular RNA content and the specific growth rate.

C. Environmental Changes

1. Temperature Changes

There are several reports on transient behavior in continuous culture following shifts in temperature. Topiwala and Sinclair[29] studied the transient response after temperature shifts in continuous culture with *Klebsiella aerogenes;* they observed a lag before the cell growth was affected and a new steady state reached. These authors developed a model to describe their experimental results. The model worked, however, only when the actual temperature was replaced with an effective temperature; the effective temperature is related to the actual culture temperature by means of a lag time and a first order time constant.

Ryu and Mateles[30] observed a similar lag between the time of a temperature shift and the time when the transient maximum or minimum growth rate is reached. After temperature shift up and shift down experiments in a nitrogen-limited chemostat with *E. coli,* the growth rate passed through maximum and minimum values, respectively, before reaching a new steady state. This is seen by the results in Figures 4 a and 4 b for a temperature shift up and down, respectively. Larger shifts resulted in greater differences between the transient maximum or minimum and the new steady-state growth rate. Further, the magnitude of the transient growth rate was less than would be expected if they followed an Arrhenius function. Both Topiwala[29] and Ryu and Mateles[30] offered explanations in terms of physiological readjustment during the lags in the transient response. Ryu and Mateles[30] proposed that only a fraction of the ribosomal RNA is involved in protein synthesis under steady-state conditions, and reserves can be activated to counteract the effect of sudden environmental changes. Further, they suggested that the ribosome content of the organism must be adjusted to compensate for the decrease or increase in ribosomal activity resulting from the shift in temperature. This concept will be discussed in detail later on.

Harder and Veldkamp[31] studying a psychrophilic *Pseudomonas* sp found that the RNA content in this organism was a function of growth temperature with a minimum at the optimum growth temperature (14°C). They considered the increased RNA content at suboptimal temperature to be a way for the organism to compensate for slower reaction rates. However, the RNA content also increased at temperatures above temperature optimum. They studied the properties of RNA at low and high temperatures by following the transient response for a chemostat to a simultaneous shift in temperature and dilution rate. Cells growing at 10 and 18°C and dilution rate 0.05 hr^{-1}, and cells growing at 14°C and dilution rate 0.10 hr^{-1} had the same RNA content. In a shift from 10°C and dilution rate 0.05 hr^{-1} to 14°C and dilution rate 0.10 hr^{-1}, the cells were able to continue growth without adjusting the cellular RNA concentration. A shift from 18°C and dilution rate 0.05 hr^{-1} to 14°C and dilution rate 0.10 hr^{-1} resulted in a drop in cell concentration during the first few hours during which the cellular RNA concentration increased. This eventually resulted in an increase in growth rates and cell concentration. RNA synthesized at 18°C, though present in sufficient quantity, is not appropriate to serve the need for higher rate of protein synthesis required at the higher growth rate. It was also shown that RNA synthesized at 18°C supported a lower rate of protein synthesis compared with RNA synthesized at 10°C.

Matsché and Andrews[32] performed transient experiments using a thermophilic *Bacillus* sp. and compared theirs and others experimental results with theoretical curves obtained using a model based on the Arrhenius relationship and with an introduction of a first order lag for the response of the growth rate. In an experiment where the

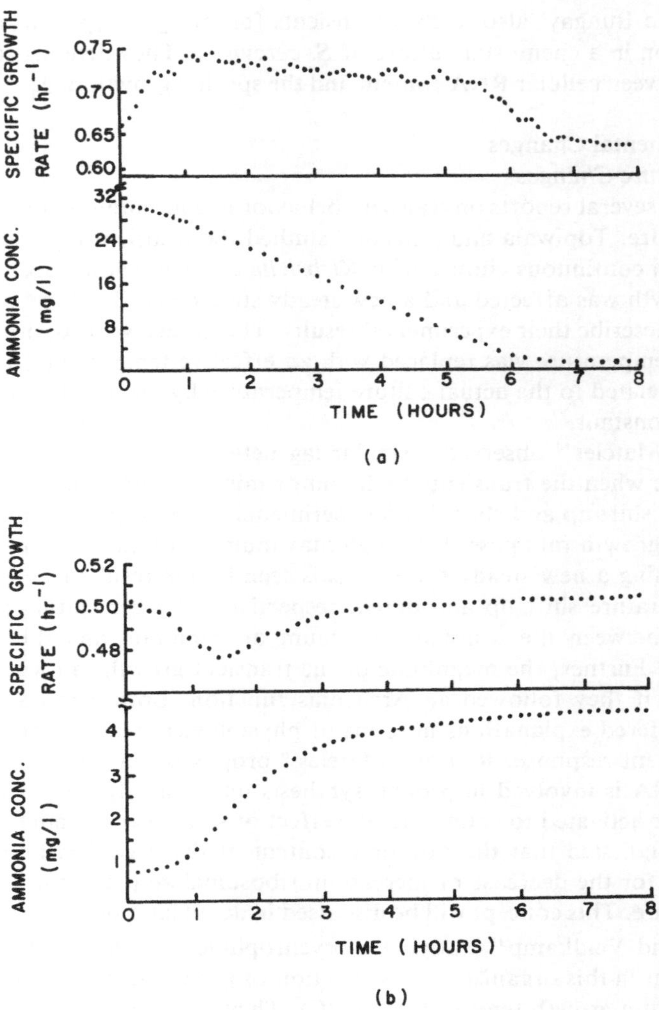

FIGURE 4. Response of *E. coli* measured by ammonia uptake during a shift-up in temperature (a) from 32 to 37°C, D = 0.64 hr,$^{-1}$ So = 75 mg/ℓ and during a shift-down (b) from 37 to 32°C, D = 0.5 hr,$^{-1}$ So = 70 mg/ℓ. (From Ryu, D. Y. and Mateles, R. I., *Biotechnol. Bioeng.*, 10, 385, 1968. With permission.)

temperature was shifted from 60 to 52°C, the cell concentration gradually increased to a new steady-state value and the simulation fitted the experimental curve quite well. When the temperature was shifted from 52 to 60°C, the model failed to predict the response. A sharp decrease in growth rate was observed accompanied by a transient accumulation of glucose. This behavior may be explained by rapid cell death and lysis of the cells. Their simulation studies using data from work by Ryu and Mateles[30] fitted fairly well to the experimental curve. Temperature shift experiments giving similar results as described above have also been carried out by Sterkin et al.[33]

2. Inhibitor Addition

There are a few reports where transient response in continuous culture has been used to study the effects of inhibitors or toxic materials on growth. Zines and Rodgers[34] studied the transient behavior of a product-limited culture in a turbidostat with *Kleb-*

siella aerogenes. The specific growth rate rapidly decreased after the pulse addition of the product inhibitor, ethanol. After the pulse, a decrease in the rate of total acid production, oxygen uptake rate, and rate of carbon dioxide release were also observed. The culture initially exhibited a similar response to step changes in ethanol concentration in the feed medium. However, the growth rate passed through a minimum and showed damped oscillations before reaching a new steady-state value. The final growth rate attained was higher than the minimum transient, and indicates adaptation of the culture to ethanol. The new steady-state values for oxygen and carbon dioxide productions were higher and the rate of acid production was lower after the step addition of ethanol. Ethanol seems to stimulate the respiratory activity and inhibit the fermentative ability of the culture. The authors also proposed a mathematical model for the transient behavior of a product-limited system.

Zines and Rodgers[35] have also applied frequency response analysis to study the effect of ethanol on the dynamic characterization of *K. aerogenes.* A nitrogen-limited chemostat was exposed to sinusoidal variations in dilution rate using a medium with or without ethanol. These experiments showed that ethanol significantly increased the time constant for the rate of total acid production and caused, in general, a slower metabolic response.

The unsteady state continuous culture was used as an experimental tool by Borzani and Vairo[36] for studying microbial inhibition. Using *S. cerevisiae,* they examined the transient response after pulse, as well as step additions of the inhibitor potassium sorbate. The specific growth rate of the yeast was drastically reduced when the inhibitor loading (gram inhibitor per gram cell) reached a critical value, regardless of the kind of disturbance used.

Lee et al.[21] examined the *Lactobacillus plantarum* growing aerobically in a glucose-limited chemostat and observed damped oscillations in cell and hydrogen peroxide concentration at low diluation rates ($D \leqslant 0.14\ hr^{-1}$) after the start of continuous flow. The period and the amplitude of the oscillations increased with decreased values of dilution rate. The production and oscillation in the concentration of hydrogen peroxide (which is known to have an inhibitory effect on lactobacilli) suggests that it may play a major role leading to the oscillations. The oscillations do not occur at high dilution rates. After a shift in dilution rate from 0.28 to 0.08 hr^{-1}, oscillations in cell and hydrogen peroxide concentration developed.

3. *Carbon Catabolite Repression*

Transient behavior in a carbon-limited chemostat with *Escherichia coli* growing on a mixture of carbon sources has been studied by Standing et al.[38] *E. coli* was exposed to a shift-up in dilution rate and substrate changes. When switched from glucose to xylose, the cell population decreased and xylose transiently accumulated in the medium before growth on xylose proceeded. Repeated shifts of the same kind showed that xylose metabolizing activity is lost during the period of growth on glucose; this behavior can be explained by catabolite repression. A preferential utilization of glucose was observed after a shift-up in dilution rate when the organism was grown on glucose and xylose or glucose and galactose. Xylose or galactose transiently accumulates in the medium after such a shift. When the culture is growing only on glucose as the carbon source, an accumulation of glucose is transiently seen in the medium after a shift up in dilution rate. When growing on a mixture of glucose and xylose or galactose, the culture is probably more capable of adapting to higher flow rates of glucose than to higher flow rates of xylose or galactose as a consequence of the constitutive nature of glucose metabolism.

4. *Aerobic/Anaerobic Transitions*

Aerobic/anaerobic transitions are difficult to review, since the literature is very empirical. This section explores effect of aerobic/anaerobic transitions on the metabolic pool sizes of ATP, NADH and related compounds, and several studies that were concerned with metabolic product formation. The literature cited is selective and hopefully is representative of a much larger body of literature concerned with effects of dissolved oxygen concentration on facultative microorganisms.

a. Metabolic Pool Size

ATP pool size and production rates have been the concern of many researchers. In an historically important study, Bauchop and Elsden[39] investigated the yield of cells per mole of ATP generated in anaerobic, glucose-limited chemostats of *Streptococcus faecalis, Saccharomyces cerevisiae,* and *Pseudomonos lindnerii.* The productivity rate of ATP was calculated by assuming a knowledge of the catabolic pathways of ATP generation. Bauchop and Elsden[39] found that the growth yield on a molar ATP basis was a constant value of about 10.5 g/mol for all the organisms tested, even though the growth yield per mole of substrate was not constant. Thus, they concluded that growth was energy dependent under these conditions.

In an effort to understand the control of ATP production in microorganisms, Cole et al.[40] measured intracellular ATP concentrations in *E. coli* and found that, in aerobic and anaerobic batch cultures of *E. coli* grown on simple medium, the ATP production rate exceeded the growth rate requirements. If *E. coli* was grown aerobically in complex medium, however, the ATP production rate was less than the growth rate requirement. The authors claimed that ATP was not limiting except in the case of complex medium. This conclusion, however, is not necessarily warranted from their observations.

Cole et al.[40] observed oscillations in the ATP pool. Holms et al.[41] took issue with this observation, claiming a constancy in the ATP pool size during batch aerobic growth of *E. coli* under which conditions pool size was only a function of carbon source.

The study by Harrison and Maitra[42] provided some interesting data regarding the effects of growth rate and dissolved oxygen level on ATP pool size and production rate. During steady-state growth in glucose-limited, aerobic cultures of *A. aerogenes,* the ATP pool was constant for growth rates exceeding 0.1 hr^{-1}. By varying the dissolved oxygen level at a growth rate of 0.18 hr^{-1}, they found that, at low dissolved oxygen levels, the ATP level declined concurrently with a decrease in the ATP/ADP ratio. The data of Haaker et al.[43] confirm these results in cultures of *A. vinelandii.*

Several of the above investigators studied the production of ATP following perturbations. Cole et al.[40] found that following a transition from aerobiosis to anaerobiosis, growth stopped briefly and the ATP level dropped rapidly, but quickly recovered to a new, lower steady-state value; subsequently, the growth rate increased. Holms et al.[41] also found an extremely rapid increase in ATP production rate following an addition of glycerol to a culture thus depleted, and suggested that mechanisms for pool maintenance are given a priority status for synthesis to enable them to respond rapidly to increases in substrate level. Harrison and Maitra[42] present similar data following a pulse of glucose to their chemostat: after an initial decrease in pool size due to utilization of ATP for biosynthesis and storage, a rapid increase in ATP productivity occurred. When their system was switched from aerobiosis to anaerobiosis, an increased flux of metabolites passed through phosphoenolpyruvate, due to what the authors claim was an increase in reaction rate "at pyruvate kinase or a stage beyond, or both".

One of the greatest difficulties in assessing the quality of data is from the experimental complexity of assaying intracellular ATP concentrations rapidly and reproducibly. The technique Cole et al.[40] used for ATP measurement was a luciferin-luciferase assay and a liquid scintillation counter; the authors claim an accuracy range of 2 to 20 pmol of ATP. The samples taken from their cultures were subjected to a 1 to 5 sec delay before quenching in perchloric acid. This delay time has received comments from various authors. Harrison and Maitra,[42] having calculated an ATP pool turnover time of about 1 sec, claim that by the time Cole et al.[40] assayed their samples they were looking at an entirely different system. Holms et al.[41] calculated an even higher pool turnover rate of 4 to 8 times per second, but claimed that, based upon aliquots taken and delayed from 3 to 30 sec before quenching in perchloric acid, there was no substantial difference in ATP concentrations.

The same, and perhaps even more difficult assay problems are encountered in analyzing the pool sizes of NAD(P)H and NAD(P). Several investigators have attempted these assays in spite of their formidable nature. Betz and Chance[44] followed the ratio of NADH to NAD and the concentrations of the three adenine nucleotides during aerobic to anaerobic shifts of *Saccharomyces carlsbergensis*. They utilized a fluorescence technique for measuring NADH and NAD. Increases in ADP and AMP essentially were in phase with increases in NADH levels. The pools of fructose-1,6-diphosphate, glycerol-6-phosphate, and dihydroxyacetone phosphate also increased in phase with the NADH levels. Conversely, the ATP pool, as one might expect, was out of phase with the other adenine nucleotides.

Wimpenny and Firth[45] also studied the relative size of the NADH to NAD pool using what appeared to be a very long, complex "cycling" assay procedure, during which time the pool sizes might change. Nevertheless, the authors provide interesting data — which reveals that the NAD (H), (i.e., NAD + NADH) pool in anaerobes is greater than the NAD(H) pool of facultative aerobes — which is, in turn, greater than the NAD(H) pool size of obligate aerobes. After a shift from aerobiosis to anaerobiosis, *Klebsiella aerogenes* showed a temporary initial increase in the NADH level for 15 min, following which time the NADH concentration returned to aerobic levels. At the same time, the NAD(H) decreased to the concentrations observed during anaerobiosis. During the same time periods, the NAD pool dropped 50 to 75% from its original value. For the same perturbation conditions, *E. coli* showed the same pattern of response, except that the response time was five times slower.

Wimpenny and Firth[45] concluded that either the NADH level may be controlled in some direct manner by the dissolved oxygen level, or internal oxidizing conditions deplete the NADH pool — or perhaps NADH concentration serves as a feedback control for its own production. They also concluded that the depletion of the NAD pool resulted from one or more of the following possibilities: a synthesis of NADH, a depletion due to the cessation of synthesis, and perhaps an oxidative deactivation of NAD.

A simple summary of the previous data that would conceptually model the changes in pool sizes of nucleotides and co-factor is improbable. The investigators, plagued by dynamic and very labile pools, have obtained mutually contradictory data. These pool sizes are no doubt of extreme relevance to regulatory mechanisms (Atkinson[46] and Harrison and Maitra[42]) and perhaps ultimately control macromolecular synthesis (Forchammer and Lindahl[47] and Maaloe[48]).

The complexity resulting from ATP pool sizes and enzyme activities was well demonstrated by Dietzler et al.[49] In this invesigation, aliquots of *E. coli* cultures that were "nitrogen starved" were subjected to various 2,4-dinitrophenol (DNP) concentrations; DNP is an uncoupler of oxidative phosphorylation. As the DNP concentration increased, both the glucose-6-phosphate (G6P) and ATP concentrations decreased. Fructose-1, 6-diphosphate concentration was independent of DNP concentration. The authors observed that the specific glucose uptake rate was independent of DNP concentration. These results, the authors claim, indicate that the regulatory link between G6P and ATP concentrations is mediated through the enzyme phosphofructokinase (PFK), an enzyme for which ATP is an allosteric effector. The control system of PFK operates such that, as ATP concentrations decrease, PFK activity increases. The authors unfortunately do not differentiate between the ATP pool size and the ATP turnover rate.

b. Product Formation

An example of the effects of dissolved oxygen on the formation of a carbon storage compound was provided by Senior et al.[50] The authors studied the accumulation of the storage compound poly-β-hydroxybutyrate (PBHB) by *Azotobacter beijerinckii* in chemostats under a variety of limitations. They found that the concentration of this intracellular product, and consequently the yield of the microorganism on the growth-limiting substrate, increased with decreasing dissolved oxygen. During transitions from nitrogen limitation to dissolved oxygen limitation, they also found that the PBHB production rate increased. They concluded that the tricarboxylic acid (TCA) cycle became limiting during anaerobiosis and that the NAD(P)H compounds produced from glycolysis were reoxidized via PBHB, rather than an electron transport mechanism.

Harrison and Pirt[51] utilized *K. aerogenes* for their aerobic to anaerobic continuous culture transitions and focused upon the end products of the fermentation pathways. They determined that the respiration rate noticeably decreased when the dissolved level fell below 10 mm Hg, at which time acetic acid production began to increase with a concurrent decrease in cell yield. Harrison and Pirt[51] also demonstrated that pH greatly altered the fermentation pathways during steady-state growth. For example, acetate production during aerobiosis and anaerobiosis at pH 7.4 greatly exceeded acetate excretion at pH 6.0, at which pH the major excretory product was ethanol. The transition phase following a shift from aerobiosis to anaerobiosis showed an extremely long time lag of 8 hr. This long lag is suggestive of the complexities of the alterations that microorganisms must make in adapting to anaerobiosis.

As can be seen from the discussion, the specific synthesis rates of excretory products are functions of medium components, pH, and dissolved oxygen (DO) concentration. Consequently, one would presume that catabolic enzymes are also influenced by the same parameters. In subsequent sections, a discussion of studies on enzyme activities and synthesis during transients is provided in order to gain some insight into the microscale effects of transient response.

D. Summary Discussion on Environmental Perturbation

Studies on transient response in continuous culture have shown that microorganisms have only a limited "reserve biosynthetic capacity" that will allow for an immediate increase in growth rate. This capacity is dependent on the nature of growth limitation,[2,3] and is restricted further under conditions of dual nutrient limitation.[5] It has been demonstrated that sudden nutritional or temperature changes may induce an oscillatory response in cell, substrate, and product concentrations with the magnitude of the shift being an important factor influencing the response pattern.

Several studies have shown that there is a close relationship between the RNA content of the cell and the specific growth rate.[31,52,53] The RNA content and its ability to support protein synthesis has also been reported to vary with temperature.[31] These results implicate ribosomal-RNA as a determinant of the biosynthetic capacity of the cell.

During transients induced by nutritional shift, changes in cell and product yield coefficients have been reported. The cell yield decreases and the yields of products of energy metabolism increase. The explanation of this response is believed to be a higher energy requirement by the cell for readjustment of its intracellular composition to meet the new growth conditions.

Transient behavior in continuous culture has further been used for studying effects of inhibitors on growth and product formation, for studying mixed cultures, and for elucidating dynamic elements in metabolic control.

Studies on the limitation of aerobic to anaerobic growth has focused strongly on the response of energy metabolism to such transition. As a consequence, considerable attention has been devoted to the difficult task of elucidating changes in co-factor pool sizes. Much work remains to be done in this area, especially in attempting to relate such transients to the problems of metabolic product formation.

III. ENZYME SYNTHESIS DURING TRANSIENT RESPONSES

As discussed previously, a transient response will, by definition, arise as a result of a perturbation occurring in a system. In the ensuing discussion of the transient response of enzyme synthesis, it will be presumed that the microbial culture had been in steady state in a chemostat prior to the perturbation and that the perturbation was a result of a rapid change in growth rate (dilution rate), dissolved oxygen concentration, nutrient concentration(s), or inducer or repressor concentration(s). In each case, the microorganism is presented with either a threat or a challenge for which a response is necessary. A summary of the enzyme synthesis response during transient in continuous culture is presented in Table 1. These results are described and discussed in the following sections.

A. Introduction

Before proceeding, the following concepts should be kept in mind: a change in in vitro enzyme activity observed during a transient response can result from an alteration in the specific enzymatic activity of extant enzyme and/or from a change in the enzyme concentration itself. Demain[54,55] discusses this phenomenon in relation to inducible/repressible enzymes, observing that the former response (inhibition or activation) is rapid, occurring in the order of seconds — whereas the latter response (induction or repression) is associated with much longer response time, on the order of minutes to hours.

The difference in response time reflects the relative complexity of subcellular events necessary to effect the changes. Inhibition or activation results from the interaction of metabolites with existing enzymes, and changes in metabolite pool concentrations occur relatively rapidly once affectors in the medium are transported into the cell. On the other hand, an enzyme concentration is the net product of the enzyme synthesis rate and the enzyme depletion rate. Enzyme synthesis is controlled by the complex, sequential, and interactive protein synthesis machinery, while the enzyme depletion rate is governed by dilution through growth and/or by selective or general proteases that are, themselves, synthesized and activated in response to perturbations. It is not surprising that this latter response is much slower and that an accurate model of the transient response of enzyme synthesis has been slow to evolve. It should also be evident from his discussion that a change in enzyme activity observed by simple measurement of in vitro enzyme activities cannot distinguish among the multitudes of possible mechanisms that can produce the observed change. Therefore, experimental results are subject to misinterpretation if these alternate mechanisms are not considered with proper experimental techniques.

B. Molecular Events Following Perturbation

Koplove and Cooney[10] have discussed selected literature that addresses the molecular biological events that ultimately express themselves in the reorganization of protein synthesis to accommodate an environmental perturbation. Although generalizations

Table 1
ENZYME SYNTHESIS DURING TRANSIENT RESPONSE IN CONTINUOUS CULTURE

Enzyme Category	Example	Perturbation	Response	Ref.
RNA polymerase	*E. coil*	Shift-up — batch fermentation		Iwakura & Ishihama[71]
Core protein			Rapid increase in synthesis rate	
σ			Independent of shift-up	
Core protein		Shift-down — batch	Differential increase in synthesis rate	
σ			Differential increase in synthesis rate	
σ′			Actively protealyzed	
β,β′ Core proteins	*E. coli*	Shift-up — batch	Rapid increase in synthesis rate	Boyle & Sells[74]
Catabolic enzyme	Histidase in *A. aerogenes*	Shift-up — chemostat	Initial rapid increase in activity, followed by slow increase in synthesis rate	Jensen & Neidhardt[89]
	Glutamic		Initial rapid increase in activity, followed by slow increase in synthesis rate	
	Dehydrogenase in *E. coli*	Shift-up — chemostat	Rapid increase in synthesis rate	Harvey[90]
	β-galactosidase in *E. coli*	Shift-up — batch	Rapid increase in synthesis rate	Dalbow & Bremer[88]
NOTE: Enzyme maintained quasi-constitutive — Expression not required for adaptive response.				
	Tryptophanase in *E. coli*	Shift-up — batch	Rapid decrease in synthesis rate	Rose & Yanofsky[57]
NOTE: Constitutive mutant — Non-essential to growth				
Enzymes of central metabolism	Acetate kinase in *E. coli*	Shift-up — chemostat	Rapid increase in synthesis rate	Koplove & Cooney[94]
Biosynthetic enzymes	Glutamic-oxalacetic transaminase in *E. coli*	Shift-up — chemostat	Slow increase in synthesis rate; comparable to increase in growth rate	Harvey[90]
	β 1,3-Glucanase in *Streptomyces* 7	Shift-down — chemostat	Rapid increase in synthesis rate	Lilley et al.[92]
	Dextransucrase in *S. sanguis*	Shift-up — chemostat	Rapid increase in synthesis rate	Carlsson & Elander[93]

are difficult, and conflicting examples can be cited, the following series of subcellular events occur in approximately the order given following a "shift up" type perturbation, (i.e., one that will permit an increase in growth rate):

1. The transcription of a region of the genome is repressed.[4,8,56,57] (This occurs in the case of nutritional enrichment.)
2. Inactive pools of ribosomes assemble into actively translating polysomes.[47,58,59]
3. Ribosomal proteins are apparently synthesized concurrently, but not necessarily coordinately with *de novo* RNA.[59,60-70]
4. RNA polymerase can be synthesized differentially and rapidly following the shift up.[71,72] Several authors believe that, under some conditions, inactive RNA pol-

ymerase exist in cells that can be activated to immediately increase the RNA synthesis rates following shift ups.[73,74]

5. The rate of RNA synthesis, specifically rRNA, increases prior to the increase in synthesis rates of protein and DNA.[75-84]
6. The synthesis rate of other enzymes and structural proteins increases.
7. The rate of DNA synthesis increases. (In special cases, in which the preshift growth rate is very slow, an immediate increase in DNA synthesis rate can occur.)[85]
8. Balanced growth at the higher growth rate is attained.

Economy is a path to survival. The microorganism that focuses its resources most efficiently and directly will have the best chance of surviving in a changing environment. Thus a hierarchy of enzymes and proteins must be established in which the most "important" ones, i.e., those whose synthesis is the sine qua non for other enzymes, are synthesized preferentially. In the above listing of subcellular responses, r-proteins and RNA polymerase are listed as being among the first respondents to shift up perturbations. Without an increase in concentration of these presumably growth-limiting proteins, there could be no increase in the net synthesis of proteins. Unfortunately, the mechanism by which these "passively controlled" operons — that is, operons for which there are no known operators such as RNA polymerase and r-protein — compete more effectively for RNA polymerase is not understood. Certain aspects of DNA topology have been suggested to be controlling mechanisms, such as the conformation of the promotor region[86] and/or the rate of unwinding and rewinding during transcription.[87] One thing has been learned by a recent study: not all nonoperator ("passively controlled") operons are synthesized at the same rate.[70] Specifically, a decontrolled lac operon was not expressed at the same rate as r-protein operons. Therefore, operon-specific topological factors may, indeed, be one factor governing metabolic regulation.

The previous discussion focused upon RNA polymerase and r-proteins because both of these classes of proteins are necessary for the synthesis of all other enzymes. Maaloe,[48] in his model of transient response, presented the concept that enzyme synthesis rates are functions of both the quantity of RNA polymerase available in the cell and the binding affinity of the promotor region of the operon — a quantity and quality concept. The remainder of this section will discuss the manner in which specific enzymes respond.

C. Enzyme Synthesis as a Means for Response

Simultaneously with the increase in RNA polymerase and r-proteins following a shift up, a variety of other transcriptional and translational events occur. Limiting or would-be limiting enzymes are selectively synthesized, while certain "unnecessary" operons are repressed. Two series of experiments prove this hypothesis unequivocally: the experiments of Dalbow and Bremer[88] and Rose and Yanofsky.[57] The experiments were similar: the former investigators maintained the inducible lactose (lac) operon in a quasi-constitutive mode with additions of cyclic AMP and isopropyl-β-D thiogalactopyranoside (IPTG), a gratuitous inducer, while in the latter, investigators selected for a constitutive tryptophan (tyrp) operon (repressor-negative). In each experimental series, the cultures were subjected to a nutritional enrichment, i.e., a shift up that resulted in an immediate increase in RNA synthesis and in an eventual increase in growth rate. However, in each case, the rate of gene expression of the lac or tryp operon decreased following the shift up. Dalbow and Bremer[88] explained their results by stating that, following a shift up, stable RNA (rRNA + RNA) is synthesized at the expense of mRNA (as Pato and von Meyenburg[73] had earlier determined) and back up their speculations with calculations. Rose and Yanofsky[57] call this complex regulatory system that governs the response of nonoperator containing genes to growth rate alterations — "metabolic regulation".

Jensen and Neidhardt[89] also studied the response of an inducible enzyme (n-histidase) during shift ups; however, they chose to maintain the operon in its wild type, inducible mode and studied the response of *Aerobacter aerogenes* during dilution rate shift ups, and the medium contained the inducer and primary carbon source histidine in excess concentration. Therefore, this experimental series differed in concept from the previously cited experiments: the conditions employed by Jensen and Neidhardt[89] required an increase in catabolic activity to occur prior to (or concurrently with) the increase in transient metabolic activity, whereas the other experimenters studied biosynthetic enzymes, which can be thought of as "secondary" or dependent enzymes, whose response reflects rather than causes an increase in metabolic activity.

Jensen and Neidhardt[89] observed results that confirmed the expected response: when the dilution rate of a chemostat at 0.13 hr^{-1} was increased to 0.66 hr^{-1}, the growth rate was observed to rapidly increase to 0.4 hr^{-1} and then, after 2 hr, to increase further to 0.66 hr^{-1}. The apparent specific synthesis rate of histidase (units per milligram cell protein-hour) increased immediately to, but did not overshoot, the post-shift apparent synthesis rate. The histidase activity (units per milligram cell protein) that is growth associated in steady state chemostats, rapidly increased following the shift and achieved steady state specific activity by the end of 2 hr.

Jensen and Neidhardt[89] thus proved rather convincingly that the activity of histidase was inhibited as well as repressed at low growth rates. Following the increase in growth rate, the inhibition was released immediately. The uninhibited histidase activity had the capacity of catabolizing histidine to meet a growth-rate demand of 0.5 hr^{-1}, which may explain the cell's ability to rapidly achieve higher growth rates with extant enzymes. In fact, they observed that cells in chemostats could adjust almost immediately with respect to growth rate and protein synthesis for small increases in dilution rates, (i.e., $D = 0.15\ hr^{-1} \rightarrow D = 0.42\ hr^{-1}$), even though histidase activity continued to increase during the first 40 min. These results are similar to those observed by other investigators.[13,80] In this case, the cell maintained a "ready reserve" of enzyme to respond to a windfall of substrate.

Harvey[90] also utilized a chemostat to study enzyme synthesis during transient response: the specific activity of glutamic dehydrogenase and glutamic-oxalacetic transaminase, growth-associated enzymes, in *E. coli,* were followed during transients initiated by the addition of glucose to glucose limited, minimal medium chemostat (Figure 5). The preshift dilution rate was 0.32 hr^{-1}, and a strain of *E. coli* that could not produce glycogen was used. Following the perturbation, the transaminase specific activity rapidly increased to the apparent unrestricted level, preceding growth rate readjustment. On the other hand, the glutamic dehydrogenase activity remained more closely coupled to the growth rate during the transient.

Glutamic dehydrogenase functions in converting ammonia to amino groups, which are then transferred into biosynthetic pathways via transamination. Having postulated that, after an addition of glucose, the growth rate of the cell apparently becomes limited by the activity of an enzyme in an amino acid biosynthetic pathway. Therefore, additions of amino acids should circumvent this metabolic block and should permit the cell to readjust much more rapidly. The same conclusion was also implied by Koch and Deppe[80] from work with a sulfate-limited chemostat, and Kennel and Magasanik[91] in studies conducted on β-galactosidase induction.

Lilley et al.[92] also were able to observe increases in enzyme synthesis rate during perturbations from steady states in continuous culture. The microorganism utilized was *Streptomyces* 17, a thermophile that produced extracellular β 1,3-glucanase. Unlike the enzymes Harvey[90] studied, the steady state profile of β 1,3-glucanase in this

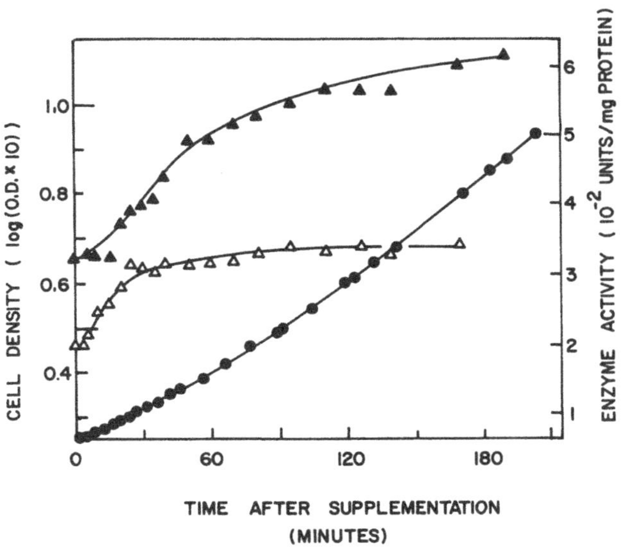

FIGURE 5. Increase in the specific activities of glucamic dehydrogenase (▲) and glutamic-oxalacetic transaminase (△, ordinate scale is 0.67×10^{-2} units/mg of protein) during the transition from restricted to unrestricted growth. At time zero, 300 m*l* were removed from a glucose-limited chemostat culture ($D = 0.32\ hr^{-1}$) of *E. coli* and added to a vessel containing 100 m*l* of basal medium. Glucose (0.05%) was added, and growth was measured by OD readings (•). See Harvey, R. J., *J. Bacteriol.*, 104, 698, 1970b for details.

microorganism was inversely proportional to growth rate, and the authors claimed this to be the result of catabolite repression.

During a shift down in dilution rate with the exclusion of the inducer gentiobiose from the medium, enzyme synthesis rates were observed to increase as much as an order of magnitude, while the specific activity increased 20 to 30% over steady state activities. The authors attributed this overshoot to catabolic derepression. However, if gentiobiose was added prior to shift down, significantly larger increases in enzyme synthesis rates and activities were obtained. The authors suggest that their observations indicate that a complicated interaction occurs between the inducer and cyclic AMP. Their results also suggest that the transient period following shift down causes a disturbance of the binding of the repressor molecules, which allows gentiobiose to act as an inducer. Of course, this hypothesis remains speculative.

Carlsson and Elander[93] studied the synthesis of the constitutive enzyme dextransucrase an extracellular polymer-forming enzyme *Streptococcus sanguis* during balanced and unbalanced growth in continuous cultures. The enzyme, which is involved in dental plaque formation, is inversely growth associated in both glucose minimal medium and complex medium. Under the former conditions, specific enzyme activity is five to sixfold less than the latter conditions. However, during a transition from the simple to the complex medium, the maximum specific dextransucrase activity during the transient can be 20 times the preshift specific activity and twice the post-shift steady state. Therefore, even though specific activity decreases with increasing growth rate in balanced growth, and the transient causes an increase in growth rate, a shift from a minimal to a complex medium caused a transient overshoot in specific activity.

Carlsson and Elander[93] presented teleological implications of their data regarding oral ecology and caries production, but do not speculate on the regulatory mechanisms responsible for this microbial response.

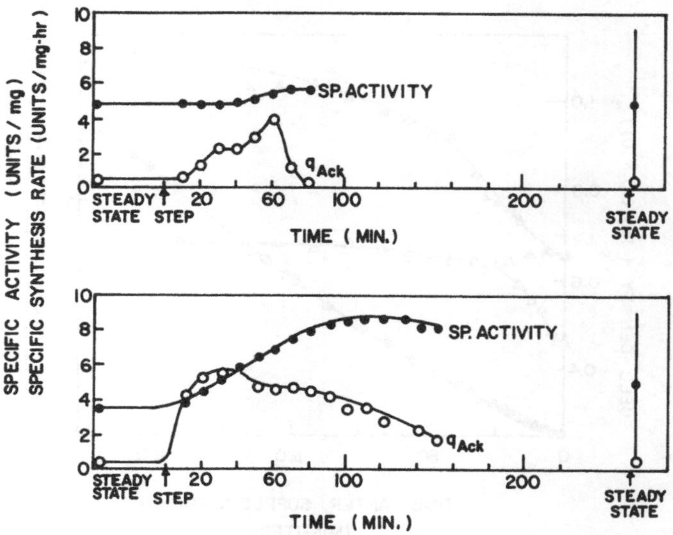

FIGURE 6. Transient response of acetate kinase synthesis during nutritional shift-up experiments with *E. coli*. The upper figure shows response to the addition of glucose. The lower curve shows the response to the addition of glucose plus casamino acids. (From Koplove, H. M. and Cooney, C. L., *J. Bacteriol.*, 134, 992, 1978. With permission.)

Relatively few experiments have focused upon the enzymes of central metabolism, such as the ATP-producing enzymes and the enzymes of the Embden-Meyerhof-Parnas pathways. One would expect that these enzymes are tightly regulated both in steady state growth and during transient responses, since their substrates and products serve as affectors for many other enzymes and an uncontrolled response could spell disaster for the cell. However, since these enzymes, especially the ATP-producing enzymes, are essential to the adaptive response of all other enzymes, one would presuppose that a complex control mechanism must exist to guarantee an "orderly" increase in transient activity. Indeed, Koplove and Cooney,[94] studying the transient response of acetate kinase in *E. coli*, observed that during nutritional shift ups in anaerobic, glucose-limited continuous cultures, marked increases (as much as fivefold) in the specific synthesis rate of acetate kinase were observed. These results illustrated in Figure 6 were explained in the context of the metabolic regulation model proposed by Rose and Yanofsky:[57] simultaneous addition of amino acids and glucose repressed the amino acid biosynthetic pathways, allowing the available *de novo* RNA polymerase to selectively transcribe unrepressed operons, such as acetate kinase. However, the experimental series was not designed to test the hypothesis. At this juncture, it appears that a major area of investigation that has not received attention is the regulation of the cellular energy balance during growth.

Many authors have attempted to model enzyme synthesis during unsteady state growth and have achieved varying degrees of success. An example of an interesting and successful model is that of Imanaka and co-workers.[95] A *Monascus* sp. was grown in continuous cultures containing glucose and galactose. High glucose concentration repressed galactose consumption and the production of α-galactosidase, and high galactose concentrations inhibited glucose utilization. The manifestation of these effects was a function of dilution rate: at $D < 0.142\ hr^{-1}$, both carbon sources were utilized; and at $D > 0.142\ hr^{-1}$, only glucose was consumed. A hysteresis effect was observed at this dilution rate: if steady states were established below $D = 0.142\ hr^{-1}$ and transients were initiated by progressively increasing the dilution rate, one response of glucose and galactose concentration, cell mass, and α - galactosidase activity was obtained. If the system was established at $D > 0.142\ hr^{-1}$, and the dilution rates were lowered rather than raised, an entirely different set of values were obtained.

A very interesting model of this system that accounted for inhibition and induction effects was presented by Imanaka and co-workers.[96] Inhibition effects were incorporated into the familiar Monod[11] model and enzyme synthesis was presumed to be a function solely of mRNA concentration, which was a function of repressor and inducer concentrations. The parameters of the model were obtained by trial and error from experimental data. Excellent agreement was obtained for chemostat cultures and for a transition between $D = 0.14\ hr^{-1}$ to $D = 0.142\ hr^{-1}$, the point of hysteresis. Imanaka et al.[97] used this model to predict the optimum reactor design for producing α - galactosidase. Therefore, in some cases that are well-defined with respect to the control of the operon, modeling is a beneficial exercise with practical benefits.

In summary, following a shift up in which a region of the genome is repressed, a preferential binding of RNA polymerase to rRNA, an r protein (and perhaps tRNA), genes occurs, which depletes the supply of polymerase available for the transcription of certain other "nonessential" genes such as lac and tryp. The enzyme products of other genes such as acetate kinase, and glutamic-oxalacetic transaminase, however, are synthesized rapidly and selectively following the shift up, perhaps indicating that these enzymes, like RNA polymerase, become bottlenecks to the metabolic synthesis network[98] and must be synthesized in the transient response.

IV. UNDERSTANDING AND APPLICATIONS OF TRANSIENT PHENOMENA

A. Introduction

By far, the major application of continuous culture has been directed towards its use in defining microbial processes under steady state conditions and in operating processes under constant environmental conditions. From the preceding discussions, however, it should be increasingly apparent that an examination of transient states in continuous culture not only provides insight into the dynamic behavior of microorganisms, but also suggests an approach towards process improvement. In the concluding section of this article, an attempt is made to identify the future trends in transient phenomena in continuous culture and to point out the major gaps in our knowledge of transient response.

B. Understanding Transient Phenomena

The intentional perturbation of a steady-state culture followed by an analysis of its response is a useful method with which to quantitate the biosynthetic capacity of the system. Close analysis of the dynamics can provide an insight into the identification of rate limiting steps. Transient response also provides an opportunity to study the sequence of key metabolic events that control a cell's response to a changing environment.

From the experiments summarized in Section III, it is apparent that the ability of a cell to respond to an environmental perturbation is dependent on the previous growth. Recently, Carter and Lorinez[99] studied the rate of protein synthesis and cell division in *S. cerevisiae* after nutritional shift ups. They found that the rate of cell division initially was maintained at the preshift value before increasing to a new and higher value. After a small shift up in growth rate (from 0.16 to 0.28^{-1}), the rate of protein synthesis changed almost immediately, whereas after a larger shift in growth rate (from 0.09 to 0.28^{-1}) the protein synthesis remained at the preshift value for 2 hr before changing. The change in the rate of cell division was also depndent on the magnitude of the shift, but always remained at the preshift value for about 2 hr after the increase in the rate of protein synthesis. Similar results were obtained by Loeb et al.[100] when studying the rate of cell division in *E. coli* B/r after nutritional shift ups. These results provide further clarification of the observations initially made by Mateles et al.[13] in 1965. It is clear that the cell has a biosynthetic capacity fixed by its previous growth rate, and that the biosynthetic capacity has some elasticity allowing for small changes in the environment. Koch[2,3] has studied in detail the kinetics of protein and nucleic acid synthesis during transients, and Koplove and Cooney[94] and Jensen and Neidhart,[89] to name examples, have studied the kinetics of enzyme synthesis during transient response to environmental perturbation; in each case, a greater understanding of the sequence of events that make up a transient response was obtained. However, the underlying control of the response still remains speculative. The rapid synthesis of RNA is essential for a major response, but what is the rate limiting mechanism(s) — i.e., is it the availability of RNA polymerase, the binding affinity for polymerase, or some as yet unidentified factor? Does the rate-limiting step change with time? Probably yes, and one is left with a gap in not only understanding the transient response, but also in devising ways to improve it.

Transient behavior is also important to elucidating the role and influence of microorganisms in natural environments. Meers and Tempest[8] examined the competition of mixed cultures in nonsteady state chemostats and showed that the method of initiating the transient effected the outcome. Our level of understanding of mixed cultures is low and much needs to be done in this area of study.

C. Application of Transient Phenomena

One important application of small volume well mixed continuous culture is in studies to investigate the effects of poor mixing in large scale fermentors. Microbial cells in poorly mixed fermentors are in a continually changing environment. By controlled manipulation of the environment in a well-mixed chemostat, it is possible to simulate the effects of poor mixing and to quantitatively evaluate the result on process performance. This approach to studying scale up provides an important test for quantifying mixing effects and also for beginning to elucidate the fundamental mechanism for the cell's response.

Discontinuities in the environment usually lead to a decrease in the efficiency of cell synthesis,[4,6,94] but an increase in the efficiency of catabolic product formation.[4,23] The reasons for this are not at all clear. However, this observation does open up the possibility of using selected perturbation on a routine basis to achieve overproduction of desired products while maximizing the conversion yield for their synthesis. It is possible to maintain cells in a continually fluctuating environment in several ways: these include the use of loop fermentors, deliberately poorly mixed fermentors, and plug flow fermentors where distance replaces time as the independent variable. The limitations in using transient phenomena to improve product formation are an understanding of the metabolic controls responsible for catabolite overproduction and a lack of knowledge in the design and operation of continuous nonsteady state fermentation.

REFERENCES

1. **Dean, A. C. R., Elwood, D. C., Evans, G. G. T., and Melling, J.**, *Continuous Culture 6*, Ellis Harwood Ltd., London, 1976.
2. **Koch, A. L.**, The adaptive responses of *Escherichia coli* to a feast or famine existence, *Adv. Microb. Physiol.*, 6, 147, 1971.
3. **Koch, A. L. and Deppe, C. S.**, In vivo assay of protein synthesizing capacity of *Escherichia coli* from slowly growing chemostat cultures, *J. Mol. Biol.*, 55, 549, 1971.
4. **Cooney, C. L., Leung, J., and Sinskey, A. J.**, Growth and physiology of *Streptococcus mutans* during transients in continuous culture, *Microbial. Abstr.*, 799—807, 1977.
5. **Young, T. B. and Bungay, H. R.**, A dynamic mathematical model of the chemostat considering mechanisms of cell growth, *Biotechnol. Bioeng.*, 15, 377, 1973 .
6. **Brooks, J. D. and Meers, J. L.**, The effect of discontinuous methanol addition on the growth of a carbon-limited culture of *Pseudomonas, J. Gen. Microbiol.*, 77, 513, 1971.
7. **Swartz, J. R.**, Computer Monitoring and Control of Continuous Culture: Yeast from Methanol, Ph.D. thesis, Massachusetts Institute of Technology, Cambridge, 1978.
8. **Meers, J. L. and Tempest, D. W.**, Influence of extracellular products on the growth of mixed microbial populations in Mg^{2+} — limited chemostat cultures, *J. Gen. Microbiol.*, 50, iv, 1968.
9. **Harrison, D. E. F. and Topiwala, H. H.**, Transient and oscillatory states of continuous culture, *Adv. Biochem. Eng.*, 3, 107, 1974.
10. **Koplove, H. M. and Cooney, C. L.**, Enzyme production during transient growth, *Adv. Biochem. Eng.*, 12, 1, 1979.
11. **Monod, J.**, The growth of bacterial cultures, *Annu. Rev. Microbiol.*, 3, 371, 1949.
12. **Monod, J.**, La technique de culture continue theorie et applications, *Ann. Inst. Pasteur (Paris)*, 79, 390, 1950.
13. **Mateles, R. I., Ryu, D. Y., and Yasuda, T.**, Measurement of unsteady-state growth rates of micro organisms, *Nature*, 208, 263, 1965.
14. **Ryu, D. Y.**, Transient Response in Continuous Culture, Ph.D. thesis, Massachusetts Institute of Technology, Cambridge, 1967.
15. **Cooney, C. L. and Wang, D. I. C.**, Transient response of *Enterobacter aerogenes* under a dual nutrient limitation in a chemostat, *Biotechnol. Bioeng.*, 18, 189, 1976.
16. **Koga, S. and Humphrey, A. E.**, Study of the dynamic behavior of the chemostat system, *Biotechnol. Bioeng.*, 9, 375, 1967.
17. **Gilley, J. W. and Bungay, H. R.**, Oscillatory growth rate responses of *Saccharomyces cerevisiae* in continuous culture, *Biotechnol. Bioeng.*, 9, 617, 1967.
18. **Mor, J. R. and Fiechter, A.**, Continuous cultivation of *Saccharomyces cerevisiae*. II. Growth on ethanol under transient-state conditions, *Biotechnol. Bioeng.*, 10, 787, 1968.
19. **Regan, D. L. and Roper, G. H.**, Response of continuous cultures to stimuli in glucose feed rate and dilution rate, *Biotechnol. Bioeng.*, 13, 815, 1971.
20. **Meers, J. L.**, Effect of dilution rate on the outcome of chemostat mixed culture experiments, *J. Gen. Microbiol.*, 67, 359, 1971.
21. **Lee, I. H., Fredrickson, A. G., and Tsuchiya, H. M.**, Dynamics of mixed culture of *Lactobacillus plantarum* and *Propionibacterium shermanii, Biotechnol. Bioeng.*, 18, 513, 1976.
22. **Cunningham, A. and Naas, P. J.**, Time lag and nutrient storage effects in the transient growth response of *Clamydomonas reinhardii* in nitrogen-limited batch and continuous culture, *J. Gen. Microbiol.*, 104, 227, 1978.
23. **Welles, J. B. and Blanch, H. W.**, The effect of discontinuous feeding on ethanol production by *Saccharomyces cerevisiae, Biotechnol. Bioeng.*, 18, 129, 1976.
24. **Vairo, M. L. R., Borzani, W., Dos Anjos, Magahaes, M. M., and Perego, L. P.**, Response of a continuous anaerobic culture to variations in the feeding rate, *Biotechnol. Bioeng.*, 19, 595, 1977.
25. **Borzani, W., Gregori, R. E., and Vairo, M. L. R.**, Response of a continuous anaerobic culture to periodic variation of the feeding mark concentration, *Biotechnol. Bioeng.*, 18, 623, 1976.
26. **Leung, J. C-Y.**, The Transient Growth of *Streptococcus mutans* in Continuous Microbial Culture, S. M. thesis, Massachusetts Institute of Technology, Cambridge, 1976.
27. **Leung, J. C-Y., Haggstrom, M., Cooney, C. L., and Sinskey, A. J.**, Steady state and transient growth of *Streptococcus mutans* in continuous culture, (submitted for publication).
28. **Luscombe, B. M.**, The effect of dropwise addition of medium on the yield of carbon-limited cultures of *Arthrobacter globiformis, J. Gen. Microbiol.*, 83, 197, 1974.
29. **Topiwala, H. H. and Sinclair, C. G.**, Temperature relationship in continuous culture, *Biotechnol. Bioeng.*, 13, 795, 1971.
30. **Ryu, D. Y. and Mateles, R. I.**, Transient response of continuous cultures to changes in temperature, *Biotechnol. Bioeng.*, 10, 385, 1968.

31. **Harder, W. and Veldkamp, H.**, Impairment of protein synthesis in an obligately psychrophilic *pseudomona* spec. grown at superoptimal temperatures, in *Continuous Cultivation of Microorganisms,* Malek, I., Beran, K., Fencl, Z., Munk, V., Ricica, J., and Smrekova, H., Eds., Academic Press, New York, 1969, 59.
32. **Matsche, N. F. and Andrews, J. F.**, A mathematical model for the continuous cultivation of thermophilic microorganisms, *Biotechnol. Bioeng. Symp.*, 4, 77, 1973.
33. **Sterkin, V. E., Chirkov, I. M., and Samoylenko, V. A.**, Study of transitional stages in continuous culture of microorganisms, *Biotechnol. Bioeng. Symp.*, 4, 53, 1973.
34. **Zines, D. O. and Rogers, P. L.**, The effect of ethanol on continuous culture stability, *Biotechnol. Bioeng.*, 12, 561, 1970.
35. **Zines, D. O. and Rogers, P. L.**, A chemostat study of ethanol inhibition, *Biotechnol. Bioeng.*, 13, 293, 1971.
36. **Borzani, W. and Vairo, M. L. R.**, Observations of continuous culture responses to additions of inhibitors, *Biotechnol. Bioeng.*, 15, 299, 1973.
37. **Lee, I. H., Fredrickson, A. G., and Tsuchiya, H. M.**, Damped oscillations in continuous culture of *Lactobacillus plantarum, J. Gen. Microbiol.*, 93, 204, 1976.
38. **Standing, C. N., Fredrickson, A. G., and Tsuchiya, H. M.**, Batch and continuous-culture transients for two substrate systems, *Appl. Microbiol.*, 23, 354, 1972.
39. **Bauchop, T. and Elsden, S. R.**, The growth of microorganisms in relation to their energy supply, *J. Gen. Microbiol.*, 23, 457, 1960.
40. **Cole, H. A., Wimpenny, J. W. T., and Hughes, D. E.**, The ATP pool in *Escherichia coli, Biochim. Biophys. Acta*, 143, 445, 1967.
41. **Holms, W. H., Hamilton, I. D., and Robertson, A. G.**, The rate of turnover of the adenosine triphosphate pool of *Escherichia coli* growing aerobically in simple defined media, *Arch. Mikrobiol.*, 83, 95, 1972.
42. **Harrison, D. E. F. and Maitra, P. K.**, "The control of respiration and metabolism in growing *Klebsiella aerogenes.* The role of adenine nucleotides", *Biochem. J.*, 112, 647, 1969.
43. **Haaker, H., Bresters, T. W., and Veeger, C.**, "Relation between anaerobic ATP synthesis from pyruvate and nitrogen fixation in *Azotobacter vinelandii", FEBS Lett.*, 23, 160, 1972.
44. **Betz, A. and Chance, B.**, "Phase relationship of glucolytic intermediates in yeast cells with oscillatory metabolic control", *Arch. Biochem. Biophys.*, 109, 585, 1965.
45. **Wimpenny, J. W. T. and Firth, A.**, "Levels of nicotinamide adenine dinucleotide and reduced nicotinamide adenine dinucleotide in facultative bacteria and the effect of oxygen", *J. Bacteriol.*, 111, 24, 1972.
46. **Atkinson, D. E.**, "The energy charge of the adenylate pool as a regulatory parameter. Interaction with feedback modifiers", *Biochemistry*, 7, 4030, 1968.
47. **Forschhammer J. and Lindahl, L.**, "Growth rate of polypeptide chains as a function of the cell growth rate in a mutant *Escherichia coli* 15", *J. Mol. Biol.*, 55, 563, 1971.
48. **Maaloe, O.**, "An analysis of bacterial growth", *Dev. Biol. Suppl.*, 3, 33, 1969.
49. **Dietzler, D. N., Peckie, M. P., and Magnani, J. L.**, "Evidence that ATP exerts control of the rate of glucose utilization in the intact *Escherichia coli* cell by altering the cellular level of glucose-6-P, an intermediate known to inhibit glucose transport *in vitro*", *Biochem. Biophys. Res. Commun.*, 60, 622, 1974.
50. **Senior, P. J., Beech, G. A., Ritchie, G. A. F., and Dawes, E. A.**, "The role of oxygen limitation in the formation of poly-β-hydroxy butyrate during batch and continuous culture of *Axotobacter beijerinckii", Biochem. J.*, 128, 1193, 1972.
51. **Harrison, D. E. F. and Pirt, S. F.**, "The influence of dissolved oxygen concentration on the respiration and glucose metabolism of *Klebsiella aerogenes* during growth", *J. Gen. Microbiol.*, 46, 193, 1967.
52. **Cooney, C. L., Wang, D. I. C., and Mateles, R. I.**, "Growth of *Enterobacter aerogenes* in chemostat with double nutrient limitations", *Appl. Environ. Microbiol.*, 31, 91, 1976.
53. **Herbert, D.**, "The chemical composition of microorganisms as a function of their environment", in Microbial Reaction to Environment, *Vol. 11, Symp. Soc. Gen. Microbiol.*, Meynell, G. G., and Gooder, H., Eds., The University Press, Cambridge, England, 1962, 391.
54. **Demain, A. L.**, "Oversynthesis of microbial enzymes", *Dev. Ind. Microbiol.*, 12, 56, 1971.
55. **Demain, A. L.**, "Cellular and environmental factors affecting the synthesis and excretion of metabolites", *J. Appl. Chem. Biotechnol.*, 22, 345, 1972.
56. **Mitsui, H., Ishihama, A., and Osawa, S.**, "Some properties of newly synthesized ribosomal ribonucleic acid in *Escherichia coli", Biochim. Biophys. Acta*, 76, 401, 1963.
57. **Rose, J. K. and Yanofsky, C.**, "Metabolic regulation of the tryptophan operon on *Escherichia coli:* repressor-independent regulation of transcription initiation frequency", *J. Mol. Biol.*, 69, 103, 1972.

58. **Alton, T. H. and Koch, A. L.**, "Unused protein synthetic capacity of *Escherichia coli* grown in phosphate-limited chemostats", *J. Mol. Biol.*, 86, 1, 1974.
59. **Harvey, R. J.**, "Regulation of ribosomal protein synthesis in *Escherichia coli*", *J. Bacteriol.*, 101, 574, 1970a.
60. **Schlief, R.**, "Control of production of ribosomal protein", *J. Mol. Biol.*, 27, 41, 1967.
61. **Stent, G.**, "Coupled regulation of bacterial RNA and protein synthesis", in *Organizational Biosynthesis*, Ghose, H. J., Lampen, J. D., and Bryson, V., Eds., Academic Press, New York, 1967, 99.
62. **Gullov, K., von Meyenburn, K., and Molin, S.**, "The size of transcriptional units for ribosomal proteins in *Escherichia coli*. Rates of synthesis of ribosomal proteins during nutritional shift-up", *Mol. Gen. Genet.*, 130, 271, 1974.
63. **Carpenter, G. and Sells, B.**, "Ribosomal and protein synthesis during a nutritional shift-up. Influence of cyclic AMP on β-galoctosidase activity", *Biochim. Biophys. Acta*, 287, 322, 1972.
64. **Carpenter, G. and Sells, B. H.**, Synthesis of individual ribosomal proteins during nutritional shift-up", *Eur. J. Biochem.*, 44, 123, 1974.
65. **Dennis, P. P.**, "Synthesis of individual ribosomal proteins in *Escherichia coli* B/r", *J. Mol. Biol.*, 89, 223, 1974.
66. **Young, R. and Dennis, P. P.**, "Balanced production of 30S and 50S ribosomal proteins after a nutritional shift-up", *J. Bacteriol.*, 124, 1618, 1975.
67. **Molin, S., von Meyenburn, K., Bullov, K., and Maaloe, O.**, "The size of transcriptional units for ribosomal proteins in *Escherichia coli*", *Mol. Gen. Genet.*, 129, 11, 1974.
68. **Bennett, P. M. and Maaloe, O.**, "The effects of fusidic acid on growth, ribosome synthesis, and RNA metabolism in *Escherichia coli*", *J. Mol. Biol.*, 90, 541, 1974.
69. **Sells, B. H., Boyle, S. M., and Carpenter, G.**, "Protein synthesis during a nutritional shift-up in *Escherichia coli*", *Biochem. Biophys. Res. Commun.*, 67, 203, 1975.
70. **Wanner, B. L., Kodaira, R., and Neidhardt, F. C.**, "Physiological regulation of a decontrolled lac operon, *J. Bacteriol.*, 130, 212, 1977.
71. **Iwakura, Y. and Ishiham, A.**, "Biosynthesis of RNA polymerase in *Escherichia coli*", *Mol. Gen. Genet.*, 142, 67, 1975.
72. **Hayward, R. S., Tittawella, I. P. B., and Scaife, J. G.**, "Evidence for specific control of RNA polymerase synthesis in *Escherichia coli*", *Nature (London) New Biol.*, 243, 6, 1973.
73. **Pato, M. L. and von Meyenburg, K.**, "Residual RNA synthesis *Escherichia coli* after inhibition of initiation of transcription by rifam picin", *Cold Springs Harbor Symp. Quant. Biol.*, 35, 497, 1970.
74. **Boyle, S. M. and Sells, B. H.**, "The relationships among RNA synthesis, RNA polymerase synthesis, and guanosine tetraphosphate levels in *Escherichia coli* during nutritional shift-up", *Arch. Biochem. Biophys.*, 172, 215, 1976.
75. **Kjeldgaard, N. O., Maaloe, O., and Schaechter, M.**, "The transition between different physiological states during balanced growth of *Salmonella typhimurium*", *J. Gen. Microbiol.*, 19, 607, 1958.
76. **Kjeldgaard, N. O. and Kurland, C. G.**, "The distribution of soluble and ribosomal RNA as a function of growth rate", *J. Mol. Biol.*, 6, 341, 1963.
77. **Neidhardt, F. C. and Fraenkel, D. G.**, "Metabolic regulation of RNA synthesis in bacteria", *Cold Springs Harbor Symp. Quant. Biol.*, 26, 63, 1961.
78. **Maaloe, O. and Kjeldgaard, N. O.**, *Control of Macromolecular Synthesis*, W. A. Benjamin, New York, 1966, 284.
79. **Koch, A. L.**, "Kinetic evidence for a nucleic acid which regulates RNA biosynthesis", *Nature (London)*, 205, 800, 1965.
80. **Koch, A. L. and Deppe, C. S.**, "*In vivo* assay of protein synthesizing capacity of *Escherichia coli* from slowly growing chemostat cultures", *J. Mol. Biol.*, 55, 549, 1971.
81. **Bremer, H. and Dennis, P. O.**, "Transition period following a nutritional shift-up in the bacterium *Escherichia coli* B/r: stable RNA and protein synthesis", *J. Theor. Biol.*, 52, 365, 1975.
82. **Nierlich, D. P.**, "Regulation of ribonucleic acid synthesis in growing bacterial cells. I. Control over the total rate of RNA synthesis", *J. Mol. Biol.*, 72, 751, 1972.
83. **Nierlich, D. P.**, "Regulation of ribonucleic acid synthesis in growing bacterial cells. II. Control over the composition of the newly made RNA", *J. Mol. Biol.*, 72, 765, 1972.
84. **Nierlich, D. P.**, "Radioactive uptake as a measurement of synthesis of messenger RNA", *Science*, 158, 1186, 1967.
85. **Sloan, J. B. and Urban, J. E.**, "Growth response of *Escherichia coli* to nutritional shift-up: immediate division stimulation in slow-growing cells", *J. Bacteriol.*, 128, 302, 1976.
86. **Sankaran, L. and Pogell, B. M.**, "Differential inhibition of catabolite-sensitive enzyme induction by intercalating dyes", *Nature (London) New Biol.*, 245, 257, 1973.
87. **Koch, A. L.**, "The adaptive responses of *Escherichia coli* to a feast and famine existence", *Adv. Microb. Physiol.*, 6, 147, 1971.

88. **Dalbow, D. G. and Bremer, H.**, "Metabolic regulation of β-galactosidase synthesis in *Escherichia coli*", *Biochem. J.*, 150, 1, 1975.
89. **Jensen, D. E. and Neidhardt, F. C.**, "Effect of growth rate on histidine catabolism and histidase synthesis in *Aerobacter aerogenes*", *J. Bacteriol.*, 98, 131, 1969.
90. **Harvey, R. J.**, "Metabolic regulation in glucose-limited chemostat cultures of *Escherichia coli*", *J. Bacteriol.*, 104, 698, 1970.
91. **Kennel, D. and Magasanik, B.**, "Control of the rate of enzyme synthesis in *Aerobacter aerogenes*", *Biochim. Biophys. Acta*, 81, 418, 1964.
92. **Lilley, G., Rowley, B. I., and Bull, A. T.**, "Exocellular β-1,3 glucanase synthesis by continuous-flow cultures of a thermophilic streptomycete", *J. Appl. Chem. Biotechnol.*, 24, 677, 1974.
93. **Carlsson, J. and Elander, B.**, "Regulation of dextransucrase formation by *Streptococcus sanguis*", *Caries Res.*, 7, 89, 1973.
94. **Koplove, H. M. and Cooney, C. L.**, "Acetate kinase production by *Escherichia coli* during steady-state and transient growth in continuous culture", *J. Bacteriol.*, 134, 992, 1978.
95. **Imanaka, T., Kaida, T., Sato, K., and Taguchi, H.**, "Optimization of α-galactosidase production by mold", *J. Ferment. Technol.*, 50, 633, 1972.
96. **Imanaka, T., Kaida, T., and Taguchi, H.** "Unsteady-state analysis of a kinetic model for cell growth and α-galactosidase production in mold", *J. Ferment. Technol.*, 51, 423, 1973.
97. **Imanaka, T., Kaieda, T., and Taguchi, H.**, "Optimization of α-galactosidase production in multi-stage continuous culture of mold", *J. Ferment. Technol.*, 51, 431, 1973.
98. **Ierusalimsky, N. D.**, "Bottle-necks in metabolism as growth rate controlling factors", in *Microbial Physiology and Continuous Culture*, Powell, E. O., Evens, C. G. T., Strange, R. E., and Tempest, D. W., Eds., Her Majesty's Stationery Office, London, 1967, 23.
99. **Carter, B. L. A. and Lorincz, A.**, Protein synthesis, cell division, and the cell cycle in *Saccharomyces cerevisiae* following a shift to a richer medium, *J. Gen. Microbiol.*, 106, 222, 1978.
100. **Loeb, A., McGrath, B. E., Navre, J. M., and Pierucci, O.**, "Cell division during nutritional upshifts of *Escherichia coli*," *J. Bacteriol.*, 136, 631, 1978.

INDEX

A

Acer pseudoplatanus
cell growth of, II: 143
chemostat studies, II: 149—150
turbidostat culture of, II: 151, 152
Acetamide limitation, of *P. aeruginosa*, II: 73
Acetate
and aerobic *E. coli* culture, II: 102
energy for transport of, II: 108
and methyl α-glucoside transport, II: 6—7
oxidation of methane to, I: 121
Acetate limitation, carbohydrate-limited growth, II: 104
Acetate kinase, I: 162, 163
Acetic acid production, I: 103, 156
α-Acetohydroxy acid synthetase, II: 70
Acetoacetyl-CoA reductase, II: 31, 33, 35
Acetobacter suboxydans, I: 146
Acetoin, I: 100, 101
Acetone, I: 1, 121
Acetyl-CoA, see also Cofactors, II: 29—30, 61
N-Acetylglucosamine, II: 50
N-Acetylmuramic acid, II: 50
Acetylornithinase, II: 131
Acids, organic, production of, I: 103
Acinetobacter, I: 118
Acridine orange, mutagenicity of, II: 133
Acridines, as antimutagens, II: 135
Adenosine, as antimutagen, II: 135
Adenylate nucleotides, dilution rate and, see also AMP; ATP, II: 79
Aeration, see also Gas(es)
in solid substrate systems, I: 94
methods for, I: 19
plant cell cultures, II: 144, 145*n*
and plant cell synchronization, II: 151, 153
Aerobacter, I: 100, 101
Aerobacter aerogens, see also *Klebsiella aerogenes*
alternate pathways of, II: 93
ammonium excess and, II: 89
ATP pools, I: 154
and Contois model, I: 44
dilution rate and enzyme synthesis in, II: 70
glycerol pathways, II: 90—91
n-histidase induction in, I: 160
parent/variant selection and, II: 130
respiratory enzymes, II: 84
special pathways of, II: 93
Aerobacter cloacae, II: 101
Aerobic/anaerobic transitions
and ATP pools, I: 154
and pool sizes, I: 155
production formation in, I: 156
and transient behavior, I: 153—154
Aerobic growth, see also Gases; Oxygen
Aerobacter cloacae, II: 102
E. coli, yield, II: 102—103
nonrespiratory oxygen utilization, II: 114—115
and oxidative phosphorylation, II: 100, 111—113
and yield shortfall, II: 111—113
Aging, see Long-term cultivation
Agglutinogens, envelope and, II: 59
Alanine, and teichoic acid, II: 50
Alcaligines megaterium, II: 19, 20
Alcohol production, I: 100, 101—102
Algae, for single cell protein production, see also Phytoplankton, I: 115
Alginic acid production, II: 56
Alkanes, and mixed cultures, I: 136
Allergens, dialysis culture of, I: 93
Amidase, II: 76
Amino acid limitation
and peptidoglycans, II: 50
and RNA translation, II: 62
S. mutans, and morphology, II: 42
Amino acids
and oxidative phosphorylation, II: 111
Pseudomonas methylotropha, pulse feeding and, I: 147
single cell protein and, I: 114
Aminoglycosides, I: 103
2-Aminopurine, II: 133, 134, 135
Ammonia, see also Nitrogen; Nitrogen limitation
excess, and glutamate dehydrogenase, II: 89
genetic engineering of transport systems, I: 136
methane mono-oxygenase oxidation by, I: 132
as nitrogen source, in methane/methanol utilizing systems, I: 131
pulse additions of to nitrogen limited culture, I: 150
and ribulose monophosphate pathway organisms, I: 129—132
transport systems for, I: 132, 135
uptake of, temperature changes and, I: 152
Ammonium limitation
and adenylate nucleotides, II: 79
B. subtilis, and envelope lipid, II: 53
and catabolic enzymes, II: 77
K. aerogenes, and slip reactions, II: 109
and lipids, II: 55
and *P. aeruginosa* glucose metabolism, II: 12—13, 82—83, 84
and peptidoglycan, II: 50, 51, 52
AMP, II: 61, 79
α-Amylase production, I: 105—106
Anabena flos-aquae, II: 167, 170
Anaerobes
ATP consumption by, II: 100
faculative, fermentative and oxidative routes in, II: 84—85
Anaerobic growth, see also Gases; Oxygen, I: 22
E. coli, yield, II: 102—103
and pullulanase production, I: 105
tubular fermenters and, I: 81—82
Anionic polymers, Gram-positive bacteria, II: 42—53

function of, II: 48
growth rate and, II: 49
ionic environment and pH effects, II: 50
lipoteichuric acids, II: 47—48
nutrient limitation and t teichoric/teichuronic acid content, II: 45—47
teichoric acid, II: 42—44, 45—47
teichuronic acid, II: 44—45, 45—47
wall turnover, II: 48—49
Ankistrodesmus facaltus, II: 175
Ankistrodesmus braunii, II: 173
Antibiotics
envelopes and sensitivity, II: 58
cross-flow system and, I: 91—92
production of, I, 1, 103—105
resistance to, selection for, II: 132—133, 137
Antifoam, I: 24, II: 2
Antigens
dialysis culture of, I: 93
envelope and, II: 59
teichoric acids as, II: 43
Antimutagens, II: 135, 138
Arachis hypogaea, II: 147
Argon, I: 22
Arrhenius function, I: 151
Arthrobacter globiformis, I: 147
Aspartic acid production, I: 103
Aspergillus Spergillus nidulans, II: 81, 84
Aspergillus ochraeus, I: 107
Asterionella formosa
cylostat growth phases, II: 164
nitrogen/phosphorus ratio for, II: 175
growth rate and subsistence quotas, II: 167
nutrient uptake, feedback and oscillation of K,, II: 170
photoperiodic activities of, II: 176
and *Scendesmus,* II: 176
ATP, see also Bioenergeics
aerobic/anaerobic transitions and, I: 154
anaerobic microorganism consumption, II: 100
carbon-energy substrates and, II: 129
dilution rate and, II: 79
and glycogen synthesis, II: 61
and maintenance metabolis, II: 117—123
maintenance demands and, II: 120—122
nitrogen source and, in methane/methanol utilizing systems, I: 131
pathways of, transient responses in, I: 162
and phosphofructokinase, II: 88
pool size, I: 154—155
slip reactions, II: 109—111
sulfate limitation and, II: 113
theoretical vs. practical values of Y_{ATP}, II: 115—117
ATPase, *E. coli* mutant, II: 116—117
Azotobacter chroococcum, II: 34
Azotobacter beijerinckii,
poly-β-hydroxybutyrate, II: 21—34
batch culture experiments, II: 21
enzyme regulation by oxygen concentration, II: 28—29
glucose metabolism enzyme regulation, II: 33—34, 35, 36
limitations affecting, I: 156
nitrogen fixing, chemostat for study of, II: 21—22
nutrient limitations and growth rate on accumulation of, II: 22—24
oxygen limitation and, II: 25—28
regulation of, II: 29—31
tricarboxylic acid cycle regulation, II: 31—33
polyphosphate in, II: 60—61
Azotobacter vinelandii, I: 154, II: 34, 56
Automation, of continuous fermentation, I: 98, 99
Auxin, II: 142

B

Bacillariophyceae, II: 177—178
Bacillus, temperature changes and, I: 151—152
Bacillus brevis, I: 104
Bacillus coagulans, I: 106
Bacillus licheniformis
anionic polymers, II: 44, 45
cell wall phosphorus, II: 46
phosphoglucomutase-deficient mutants, II: 47
Bacillus megaterium
carbon and energy reserves of, II: 17
and cell wall phosphorus, II: 54
DNA and protein contents of, II: 62
poly-β-hydroxybutyrate in, II: 19, 20
polyphosphate of, II: 61
terramycin resistant, II: 132—133
Bacillus subtilis
amylase production by, I: 105
cell wall, II: 46, 47
envelope composition of, II: 45, 50—53
in mixed culture, I: 146
mutation selection, II: 137
nutrient limitation and, II: 40-41, 46, 47
peptidoglycans of, II: 50—53
protease production by, I: 105
teichoic acids of, II: 42—43, 44
teichoic/teichuronic acids of, II: 50
wall turnover in, II: 48—49
Bacillus stearothermophilus, I: 105
Bacteria, see also Species
dilution rate and enzyme synthesis in, II: 70—79
morphology, II: 46
photosynthetic purple sulfur, II: 176
single cell protein production, I: 115
Bacteriophage
binding sites, II: 48—49, 55
resistance to, II: 133, 137
Base pair transitions, II: 138
Batch culture
B. subtilis, and envelope lipids, II: 53
vs. continuous culture, I: 8—9
enrichment for hydrocarbon utilizing organisms in, I: 116

extracellular glucose oxidation in, II: 15
fumaric acid, I: 103
growth data in, I: 64—65, 98
Monod growth kinetics, I: 34—35
penicillin production, I: 104
product formation kinetics, I: 77—78
product induction in, I: 106
S. mutans, glycogen synthesis, II: 61
Bdellovibrio bacteriovorus, II: 115
Beer production, I: 107—108
Benzylformate limitation, II: 130
Bioconversion processes, I: 106—107
Biochemical Oxygen Demand (BOD), I: 109
Biochemistry, see also Biosynthesis; Enzymes, specific pathways
in single cell protein production, I: 134—135
intracellular, environmental alterations and, I: 100
methane/methanol utilization, I: 130—131
transients and, I: 157—159
Bioenergetics
maintenance metabolism, II: 117—123
population density determinants, II: 101—102
practical considerations, II: 102—104
substrate dependency and molar growth yield, II: 104—109
theoretical vs. practical values of Y_{ATP}, II: 115—117
yield shortball, II: 109—115
Biofeedback systems, I: 88—92
Biomass, see also Cell concentration; Cell moss; Population density
determinants of, theory, II: 101
dilution rate and, in gas-limited culture, I: 123, 124
and equations of balance, I: 28, 29
increasing, I: 107
in nutrient-limited conditions, I: 21
in single cell protein production, mixed vs. pure cultures, I: 119—120
nutrient concentration vs., I: 39
nutrient limitation, and dilution rate, II: 40—41
phytoplankton, II: 166—169
plant cell chemostat, II: 148, 149
protein content of, carbon-energy limitation and, I: 124
selection, and loss of, I: 23
wall growth, I: 55—56
with methane/methanol substrate, I: 122—124
Biopolymers, see also Glycogen; Poly-β-hydroxybutrate; Polyphosphate
ATP consumption for, II: 100, 104
energy reserve, II: 60—61
envelope, see Envelope
leakage of, II: 119—120
Biosynthesis, see also Biochemistry
coexistence and, II: 176
dilution rate and, II: 70
reserve capacity, I: 156
Blue-green alage, I: 114
Bordetella pertussis, II: 59
Budding, I: 146, 149
Butanol production, I: 1
Butanol-acetone fermentation, I: 101—102
2, 3 Butylene glycol production, I: 100

C

Cachonia niei, II: 179
Cadmium, II: 58
Caffeine, II: 133, 134—135
Callus formation, II: 142
Calvin cycle, I: 129
Candida utilis, II: 81
Carbohydrate metabolism
Acer pseudoplatanus, II: 149, 150
in *P. aeruginosa*, II: 3—16
carbon limitation and, II: 14—16
glucose and diauxic growth, II: 3—5
glucose, gluconate, and 2-oxogluconate transport, II: 11—14
glucose transport, II: 6—11
Carbohydrates
and ATP consumption, II: 100
budding and, I: 149
growth rate and, I: 7
phytoplankton, circadian rhythms, II: 174
polymers of, in envelopes, II: 42—43
for single cell protein production, I: 114
transients and consumption of, I: 162—163
Carbon
catabolite repression, I: 153
excretion of, and single cell protein production, I: 118
organic
in effluent water, I: 135
in SCP production systems, I: 121
photoperiod and, in phytoplanton, II: 174
Carbon dioxide, see also Gases
and cell carbon, in pseudomonas C, I: 136
and hydrocarbon-utilizing organisms, I: 116
and penicillin production, I: 104
Pseudomonas methylotropha, pulse feeding and, I: 147
and plant cell synchronization, II: 151, 153
pressure and, fermenter design and, I: 133—134
and product formation, in complex systems, I: 79
serine pathway organisms and, I: 130
and substrate-controlled cultures, I: 16
Carbon-energy, see Methane; methanol
Carbon limitation, see also Nutrient limitation
Azotobacter beijerinckii,
and product formation, II: 22, 23
and oxygen, II: 25
and carbohydrate metabolism in *P. aeruginosa*, II: 14—16
and cryoresistance, I: 4, 5, 6
dilution rate and enzyme synthesis, II: 70
E. coli, changing carbon source, II: 102—103
and envelope proteins, II: 55

and extracellular carbon, II: 129—130
K. aerogenes, II: 109
Pseudomonas sp., and catabolic enzymes, II: 76—77
pulse feeding in, I; 147, 149
and slip reactions, II: 109, 111
Carbon metabolism
carbohydrate, in *P. aeruginosa,* II: 3—16
carbon limitation and, II: 14—16
citrate and, II: 6
glucose, gluconate, and 2-oxogluconate, II: 11—14
glucose metabolism and diauxic growth, II: 3—5
carbon and energy reserve compounds, II: 16—34
glycogen and glycogen-like polymers, II: 19
poly-β-hydroxybutyrate, I: 19-34
continuous cultivation and, II: 2
methodology, II: 2—3
objectives, II: 2
Thiobacillus novellus, II: 88—89, 90, 92
Carbon reserve compounds
glycogen and glycogen-like compounds, II: 19
poly-β-hydroxybutyrate
in *Alcaligenes eutrophus,* II: 19—21
in *Azotobacter beijerinckii,* II: 21—34
in *Bacillus megaterium,* II: 19, 20
Carbon sources, see also Specific carbon sources
and carbohydrate consumption, I: 162—163
and carbohydrates metabolism, in *P. aeruginosa,* II: 3—4
dilution rate and enzyme synthesis, II: 71—72
and *E. coli* yield, II: 102—103
methanol as, see Methanol; Methane/methanol
and oxidative phosphorylation, II: 111
and superoxide dismutase, II: 114
and transport systems, in *P. aeruginosa,* II: 9
Catabolic derepression, I: 161
Catabolite repression, see also Inhibition
carbon, I: 153
in enzyme synthesis, II: 73—75
in single stream multi-stage system, I: 73
and oxidative phosphorylation, II: 111
Catharanthus roseus, II: 143
Cations
osmotic regulation and, II: 119
teichoic acids and, II: 45, 50
Cell concentration, see also Biomass; Cell mass; Population density; Yield and foaming, I: 24
in dialysis fermentation, I: 93
plant cell chemostat, II: 148
Cell cycle, and morphology, II: 41—42
Cell division, see also Generation time; Synchronization nutritional shift-ups, and I: 163—164
phytoplankton, II: 177—178
and coexistence, II: 175
light/dark cycle and, II: 174
timing of, II: 177—178
Cell mass, see also Biomass; Cell concentration; Population density, Yield
in complex systems
and product formation, I: 77, 78, 79
substrate utilization and, I: 74
in single-stream multi-stage systems, I: 65—71
Cell quota, phytoplankton, II: 166—169, see also Biomass; Cell concentration; Yield
Cells
collection of, I: 20
immobilization of, and productivity, I: 109
Cellulase production, I: 87—88
Cellulomonas, I: 107
Cellulose
multi-stage fermentation of, I: 87—88
solid-state fermentation, I: 93—94
Ceratium dens, II: 179
Ceratium furca, II: 174, 179
Chaetoceros gracilis
cell division, II: 179
cyclostat growth phases, II: 164
growth rate and subsistence quotas, II: 167
Chelators
and *Pseudomonas* EN yield, I: 120—121
resistance to, II: 58
Chemical Oxygen Demand (COD), I: 109
Chemicals
bioconversion of, I: 106—107
toxic impurities in, I: 133
Chemiosmotic hypothesis, II: 100
Chemostats
in carbon metabolism studies, II: 2
commercially available, I: 23, 25
conversion to turbidostat, I: 20
defined, I: 14, 16
discontinous, mutation repair in, II: 135
dynamics close to equilibrium, I: 35—39
extensions of basic theory, I: 46—60
chemical inhibition, I: 47—52
heterogeneous conditions, I: 55—60
multiple species, I: 53—55
multiple substrate systems, I: 52—53
temperature variations, I: 46—47
for pathogens, I: 23
Monod growth kinetics, I: 31—32
nutrient-limitation in, I: 21
phytoplankton
algae, II: 160—161
growth and cell quota, II: 166—167
growth rate and nutrient concentration, II: 162
nutrient pools and biosynthesis, II: 173
plant cells, II: 144—153
closed continuous system, II: 148
open continuous system, II: 148—151
semicontinuous systems, II: 146
synchronous system, II: 151, 152, 153
transient behavior, theory, I: 39—42
selection in, II: 127
Chlamydomonas moewusii, II; 179
Chlamydomonas reinhardii, I: 146; II; 178
Chloramphenicol, I: 147, II: 4
Chlorella, II: 174

Chlorella elipsoidea, II: 173
Chlorella fusca, II: 179
Chlorella pyrenoidosa, II: 173
Chloretetracycline, I: 104
Chromatium vinosum, II: 176—177
Chromatium vessei, II: 176—177
Circadium rhythms, see Light; Photopreiod
Citrate
and carbohydrate metabolism in *P. aeruginosa,* II: 3, 4, 7, 9, 86
and extracellular glucose metabolism, II: 14
and intracellular glucose metabolism, II: 15
and transport systems, II: 6, 9, 11, 12, 16, 83, 84
and *Pseudomonas* EN *yield,* I: 120—121
Citrate synthetase, II: 30, 33
Citrus waste, I: 101
Closed continuous culture systems, plant cell, II: 148
"C" mode, II: 59, 60
Cobalt, I: 100
Coccal form, growth rate and, II: 40
Coccolithus huxleyi
carbon assimilation, II: 174
cell division, II: 177, 179
Cofactors, see also NAD cofactors
Azotobacter beijerinckii
and poly-β-hydroxybutyrate regulation, II: 29—31
oxygen and, II: 28—29
dilution rate and, II: 78—79
and enzyme induction, II: 130
in enzyme synthesis, II: 78—79
and polyphosphate, II: 60—61
poly-β-hydroxybutyric acid and, I: 156; II: 29—31
pool sizes, I: 155, 157
Colicins, II: 55, 133
Columns, packed, and culture dynamics, I: 56—60
Competition, see also Selection
continuous culture and, I: 8, 9
and evolution, II: 136—139
phytoplankton, II: 175—177
transients and, I: 164
Complex systems
multi-stream systems, I: 87—94
dialysis fermentation, I: 92—93
microorganism inflow, I: 88—92
solid substrate fermentation, I: 93—94
substrate inflow, I: 87—88
single-stream gradient, I: 81—87
simulation of plug-flow fermentor, I: 84—87
tubular fermentor, I: 81—84
single stream multi-stage, I: 64—81
growth characteristics, I: 64—75
product formation, I: 75—81
Contamination
during innoculation, I: 21, 99
external, I: 23
in industrial processes, I: 99
internal, I: 23
of inflow medium, I: 18
selection and, II: 128
Continuous culture systems
advances in, I: 1
apparatus, commercially available, I: 17, 23
basic concepts, I: 14, 15, 16, 17
complications of, I: 23—25
developments in, I: 1—10
holding times in, I: 69, 70—71
operation, I: 20—22
specialized situations, I: 22—23
types of, I: 17—20
Contois model, I: 43—44
Cooling
carbon-energy substrate systems, I: 128
fermenter design and, I: 133—134
and quality control, I: 99
Copper, I: 100
Corn-steep liquor, I: 103
Corynebacterium glutamicum, I: 103
Cross-flow systems, I: 91—92
Cryoresistance, I: 4, 5, 6
Cultrate 6300 Fermentation Substrate, I: 104
Culture units, plant cell, II: 144—145
Curvularia lunata, I: 107
Cyanic mode, II: 59, 60
Cyclic AMP
E. coli
and catabolite repressible enzymes, II: 73
and β-galactosidase, II: 75—76
and induction, I: 159, 161, II: 130
and oxidative phosphorylation, I: 111—112
Cyclostat, phytoplankton
algae, II: 161—162
coexistence in, II: 175—176
growth and cell quota, II: 166—169
growth rate and nutrient concentration, II: 162—166
nutrient pools and biosynthesis, II: 173—175
Cyclotella meneghiniana, II: 167
Cytrochrome oxidase
in methane/methanol utilization, I: 130—131
and oxidative phosphorylation, II: 122
Cytoplasmic membrane, see also Envelope, II: 42, 53

D

Daucus carota, II: 143
2-Deoxyglucose, II: 6
Depolymerizing enzymes, poly-β-hydroxybutyrate, II: 30
Depression, and hyperproduction, see also DNA, II: 129—130.
Desulfovibrio, II: 100
Dextransucrase, I: 161
DL-1, I: 120
Dialysis fermentation, I: 92—93
Diatoma
cell division, II: 177—178
polyphosphates in, II: 173

Diatoma elongatum, II: 167
Diauxic growth, glucose metabolism and, II: 3—5
2, 4-Dichlorophenoxyacetic acid (2, 4-D), II: 142
Differentiation, phytoplankton, II: 161—162
Diffusion-limited growth, I: 60
Digitalis lanata, II: 143
Dilution rate, see also Nutrient-limited growth; Theory
  Azotobacter beijerinckii, and product formation, II: 22, 23
  biofeedback systems, I: 89—91
  and biomass
    nutritional limitations and, II: 40—41
    theory, II: 101
  and carbohydrate metabolism, in *P. aeruginosa*, II: 5
  and enzyme synthesis
  in bacteria, I: 160; II: 70—79
    in eukaryotes, II: 79—81
    and extracellular carbon, II: 130
  gaseous carbon-energy substrate, I: 124
  and growth rate
    enzyme induction, I: 160, 161
    in steady state conditions, I: 14, 30
  and growth rate
    in steady state conditions, I: 30
    with methane/methanol substrate, I: 117, 121—123
  in multi-stage systems, I: 68, 73
  in multi-stage vs. single-stage systems, I: 72
  in single-stream multi-stage systems, I: 65—66, 68
  and metabolism, II: 3
  and *Methylomonas methanolica* productivity, I: 118
  phytoplankton, II: 160—161
  plant cell chemostat, II: 148
  product-controlled, I: 16
  and respiration, II: 110
  and ribulose monophosphate vs. serine pathway organisms, I: 122
  and selection, II: 127
  and selection for hydrocarbon utilizing organisms, I: 117
  substrate affinity and, I: 125—126
  and transient behavior, I: 145—146, 161
  viability and, I: 8
Dinitrogen, I: 131, 136
2, 4 Dinitrophenol, I: 155
Dinophysis fortii, II: 179
Diplococcus pneumoniae, II: 43
Disc feeder, volumetric, I: 87
Ditylum brightwellii
  carbon assimilation, II: 174
  cell division, II: 177, 178
  cell quota oscillations, II: 168
  cyclostat growth phases, II: 164
  photoperiodic activities of, II: 174, 176, 177
DNA, see also Genes
  delection mutants, II: 137
  growth rate and II: 62
  mutagens and, II: 134—135
  phytoplankton, circadian rhythms, II: 175
  Q-cells, II: 143
  repair of, II: 135, 138
  substrate and, in multi-stream system, I: 88
  substrate limitation and, I: 73
  synthesis of, I: 8—9, 73
DNA polymerase, antimutator and, II: 138
DNA replication
  and mutation, II: 133
  in phytoplankton, II: 173
DNA transcription
  and hyperproduction, II: 129
  transients and, I: 158—160, 163
Dreschel bottle, I: 18
Droop's equation, II: 166, 167, 168, 169, 173
Dunaliella tertiolecta
  carbon assimilation, II: 174
  cell division, II: 177, 179
  growth rate and subsistence quota, II: 167

E

Ecological problems, in chemostat, see also Complex systems; Mixed culture, Phytoplankton; Selection, I: 8, 9
  and transient phenomena
    aerobic/anaerobic transitons, I: 153—156
    carbon catabolite repression, I: 153
    inhibitor addition and, I: 152—153
    nutritional changes and, I: 145—151
  and wall growth, I: 24
Enzyme induction
  dilution rate and, II: 73—75
  and genetic selection, II: 130
  and glucose metabolism enzyme activity, II: 3
  starvation and, II: 130
  transients and, I: 157—163
Enzymes, see also Respiratory enzymes; specific enzymes and pathways
  dialysis culture and, I: 93
  dispensible, selection for, II: 130—131, 137
  extracellular, see Extracellular pathways
    industrial production of, I: 105—106
Enzyme synthesis, see also specific pathways
  adaptive significance, II: 77—78
  dilution rate and, II: 70—81
  dilution rate and
    in bacteria, II: 70—79
    in eukaryotes, II: 79—81
  isoenzymes, II: 89—94
  metabolites and, II: 75—77
  mixed substrate utilization, II: 88—89
  mutations of, II: 137
  mutations of, II: 137
  special pathways, II: 89—94
  substrate concentration and, II: 81—88
    fermentation and oxidative routes in facultative
    anaerobes, II: 84—85
    glucose metabolism, in *P. aeruginosa*, II: 82—83, 84

Pasteur effect, II: 85—88
transient responses and, I: 157—163
Equations of balance, see also Theory, I: 28—30, 58
Erwinia amylovora, II: 70, 84
Erythromycin production, I: 104—105
Escherichia coli, see also Bioenergetics
acetate kinase, transient responses in, I: 162
anerobic vs. aerobic growth, II: 102—103
ATPaseless mutant, II: 116—117
ATP pool size, I: 154
carbon catabolite repression in, I: 153; I: 73
carbon limitation
and catabolite repressible enzymes, II: 73
and β-galactosidase, II: 75
cell division in, I: 164
cofactor pool sizes, I: 155
cryoresistance of, I: 5, 6
dilution rate changes, and transient behavior, I: 145
enzyme production, I: 106
enzyme induction, II: 130—131
enzyme induction during transient response, I: 160
β-galactosidase
Economics, of single cell protein production, I: 122—123, 134—136
Economy, metabolic
and selection, II: 137
transients and, I: 159
Ecostat, I: 91
EDTA, resistance to, II: 58
Effluents, see also Industrial wastes; Waste waters
collection of, I: 20
organic carbon in, I: 136
oxygen demand reduction for, I: 109
recycling of, methanol-utilizing systems, I: 124
sampling from, I: 22
Electrical failure, I: 25
Electrolytes, osmotic regulation and, II: 199
Embden-Meyerhoff-Parnas pathway
and ATP consumption, II: 100
in *Acer pseudoplatanus*, II: 149, 150
oxygen and, II: 85
Pasteur effect and, II: 85, 88
transient responses in, I: 162
Energy, see also Bioenergetics
budding and, I: 147
for dinitrogen fixation, I: 136
photosynthesis and,II: 174
spillage, II: 109—111, 116
Energy limitation, see Methane/methanol substrates
Energy reserve compounds, II: 60—61
glycogen and glycogen-like compounds, II: 19
poly-β-hydroxybutyrate
in *Alcaligenes eutrophus*, II: 19—21
in *Azotobacter beijerinckii*, II: 21—34
in *Bacillus megaterium*, II: 19, 20
Enrichment culture, see also Selection
for genetic selection, II: 127
for hydrocarbon utilizing organisms, I: 116—117
Enterobacter aerogenes, I: 145, 147—148
Entner-Doudorff enzymes
Azotobacter beijerinckii
nutrient limitation transitions and, II: 34
oxygen and, II: 28—29, 35
citrate/glucose ratio and, II: 86
P. aeruginosa, carbon limitation and, II: 14
Envelopes
biological properties of, II: 58—60
cytoplasmic membrane, II: 53
exopolysaccharides, II: 55—57
Gram-negative bacteria, II: 53—55
Gram-positive bacteria, II: 42—53
anionic polymer function, II: 48
anionic polymers, II: 42—50
growth rate and, II: 49
ionic environment and pH effects, II: 50
lipoteichoic acids, II: 47—48
nutrient limitation and teichoic/teichuronic acid content, II: 45—47
teichoic acid in, II: 42—44, 45—47
teichuronic acid in, II: 42—45, 45—47
wall turnover, II: 48—49
intracellular macromolecules, II: 60—63
leakage, II: 119—120
peptidoglycans and, II: 50—53
P. aeruginosa, II: 59
Environment, see also Growth conditions; Transients for enrichment for hydrocarbon utilizing organisms, I: 116—117
and equations of balance, I: 29
and intracellular biochemistry, I: 100
and Monod model, I: 43—44
and physiology, I: 2, 3—4
carbon limitation and, II: 75
hyper levels of, II: 129, 136
glutamate dehydrogenase pathway, I: 136
glycogen in, II: 19, 20, 21, 61
K100 antigen of, II: 43
maintenance metabolism of, II: 120—121, 122
membrane proteins, iron and, II: 55
mixed substrate utilization, II: 88, 89
mutants, membrane protein, II: 54
mutant selection, II: 128, 132, 133, 134, 136, 138
mutation repair in, II: 135, 136
neomycin resistant, selection for, II: 132
nutritional shift-ups and, I: 149, 164
oxidative phosphorylation, II: 100, 111—112
peptidoglygan turnover, II: 118
phage resistance of, II: 133
plasmid stability, II: 62—63
RNA regulation in, II: 62
slip reactions, II: 111
temperature changes and, I: 151, 152
tricarboxylic acid cycle, II: 33
trp, reversion of, II: 131—132
yield, determinants of, II: 102—103
Ethane
and dilution rate, in methane-limitation, I: 126

and mixed cultures, I: 136
and single cell protein production, I: 121
Ethanol
in butylene glycol fermentation, I: 100, 101
as product inhibitor, in *K. aerogenes*, I: 152—153
production of, I: 1, 146
for single cell protein production, I: 115
Ethanolamine, II: 58
Ethylene
for single cell protein production, I: 115
synchronizing with, plant cells, II: 151, 154
Ethylene oxide sterilization, I: 21
Euglena gracilis
cell division, II: 177, 178, 179
cell quota oscillations, II: 168, 169
circadian rhythm, II: 174
cyclostat growth phases, II: 164
growth rate and subsistance quota, II: 167
nutrient limitation, I: 162—163
nutrient uptake oscillations, II: 171 173
nutrient uptake, feedback and oscillation of K., II: 170, 171, 172, 173
phosphate uptake, II: 174
photoperiodic activities, II: 176
polyphosphate pools, II: 173, 174
Eukaryotes
enzyme synthesis in, II: 79—81
minimum growth rate, I: 7
Evaporation, prevention of, I: 18
Evolution
continuous culture and, II: 136—139
plant cell cultures, II: 143
Exopolysaccharides, II: 55—57
Exponential growth
in closed continuous system, II: 148
defined, I: 2—3
extending, I: 107
Extracellular nutrients, phytoplankton, II: 173
Extracellular pathways
glucose metabolism, in *P. aeruginosa*, II: 12, 14—16, 84, 91
and supernatant carbon, in SCP production, I: 119

F

Fatty acids, II: 53, 118
Feedbeck systems, I: 88—92
Fermentation, see also Aerobic/anaerobic transitions; Anaerobic growth
dialysis, I: 92—93
glucose, ATP production in, II: 100
fixed matrix, I: 102, 109
routes of, in facultative anaerobes, II: 84—85
solid substrate, I: 93—94
and superoxide dismutase, II: 114
Fermentation industry
beer production, continuous, I: 107—108
development of, I: 108—109
processes available for development, I: 100—107
growth associated products, I: 100—103
non-growth associated products, I: 103—106
transformation products, I: 106—107
Fermenter
I.C.I. pressure cycle, I: 136
plant cell, II: 144—153
plug-flow, I: 81—84
plug flow, simulation of, I: 84—87
in single cell protein production, I: 133—134
tubular, I: 81—84
Filamentous organisms
long-term cultivation of, I: 73
for SCP production, I: 115
Fixed matrix fermentations, I: 102, 109
Flavobacterium, I: 118
Flow rate
and dilution rate, I: 14
methods of controlling, I: 17—18
Foaming, see Antifoam
Fodder, fermentation of, I: 93
Formaldehyde dehydrogenase, I: 130
Formate
and aerobic *E. coli* culture, II: 102
B. oxalaticus and, II: 89
and yield shortfall, II: 115
Formate dehydrogenase, I: 130
Freezing
cryoresistance, I:4
plant cells, II: 143
Fucose, and methyl α-glucoside, see also Carbon sources, II: 6
Fumaric acid, production of, I: 103
Fungi, filamentous, I: 73, 115

G

Galactose, see also Carbon source
E. coli, accumulation of, I: 153
and methyl α-glucoside, II: 6
utilization of, in *Monascus*, I: 162—163
α-Galactosidase, production of, I: 106, 162—163
β-Galactosidase
cAMP and, II: 73, 75—76
E. coli
"hyper" mutants of, II: 77—78, 129, 136
induction and catabolite repression of, I: 106; II: 75
enteric bacteria and, I: 5
K. aerogenes, II: 131
Gamma radiation, II: 133
Gas(es), see also Aeration; Carbon dioxide; Oxygen dilution and, II: 3
fermenter design and, I: 135
in plant cell synchronization, II: 151, 153
in simulated plug-flow fermenter, I: 86
in tower fermenter, I: 87
in tubular fermenter, I: 82
and product formation, in complex systems, I:

79
Gas-limited cultures
biomass in, dilution rate and, I: 122, 123
steady state biomass and dilution rate in, I: 124
Generation time, mutagens and, see also Cell division, II: 133, 134
Genes, see also DNA; Selection
hyperproducing, II: 77—78, 129, 136
nonessential
selection and, II: 130—131, 137
transients and, I: 163
plant cell, II: 143
rel A, and ribosomal RNA synthesis, II: 62
transients and expression of, I: 158—159, 159—160, 163
Treffer's mutation, II: 138
Genetic engineering, and methanol utilizing bacteria, I: 136
Genetic studies, II: 127—139
evolution II: 136—137
factors affecting selection, II: 127—128
mutation process, II: 133—136
parent-variation selection, II: 128—133
Gentobiose, I: 161
Glenodinium, II: 174
Glossary, phytoplankton, II: 181
β 1, 3 Glucanase, induction, I: 160—161
Gluconate, in *P. aeruginosa*
and extracellular pathways, II: 15—16
glucose limitation and, II: 14
and transport systems, II: 6—7, 9, 11—14, 17
uptake and utilization of, II: 10, 11
Gluconate dehydrogenase, II: 86
Glucose
accumulation of, and growth rate, I: 152
E. coli, preferential utilization of, I: 153
and poly-β-hydroxybutyrate, II: 21
Glucosamine, II: 6, 10
Glucose/citrate ratio, and enzyme activity, II: 86
Glucose isomerase, I: 106
Glucose limitation
and envelope composition, of *P. aeruginosa*, II: 59
glucose addition and respiration, II: 110
and lipids, II: 55
and lipoteichoic acids, II: 48
mutation repair in, II: 135, 136
and oxidative phosphorylation, II: 111—112
and peptidoglycans, II: 51, 53
and plasmid stability, II: 62
and polysaccharide formation, II: 56
pulse feeding in, I: 147
and xanthan production, II: 57
Glucose metabolism
Azotobacter beijerinckii
regulation of, II: 33—34, 35, 36
nutrient limitation and, II: 32, 34
oxygen and, II: 28
and diauxic growth, II: 3—5
mixed substrates and, II: 17, 86, 89
oxygen and, II: 87
P. aeruginosa, II: 6—11, 11—14, 82—83, 84, 91
citrate and, II: 6, see also Citrate
nutrient limitation and, II: 14, 18
pathways of, II: 14, 15, 91
transport systems, II: 9, 11—14, 16, 17
uptake and utilization of, II: 10
Glucose-6-phosphate, I: 155
Glucose-6-phosphate dehydrogenase, II: 35, 86
Glucosylglycerol phosphate teichoic acid, II: 44
Glutamate dehydrogenase
ammonium excess and, II: 89
E. coli
induction of, I: 160, 161
replacement of ATP dependent GS pathway with, I: 136
Glutamate synthetase, ATP dependent, I: 132, 136
Glutamic acid, production, I: 103
Glutamine-oxoglutarate transaminase
ammonium limitation and, II: 89
induction of, in *E. coli*, I; 160, 161
synthesis of, II: 70
transients and, I: 163
Glutamine synthetase
ammonium excess and, II: 89
ATP dependent, I: 132, 136
D-Glyceraldehyde-3-phosphate dehydrogenase, II: 19, 70
Glycerol
Aerobacter aerogenes pathways, II: 90—91
P. aeruginosa carbohydrate metabolism, II: 3, 9, 11, 12
Glycerol dehydrogenase, II: 70
Glycerol limitation, see also Carbon limitation
E. coli
and β-galactosidase, II: 75
and yield shortfall, II: 113
K. aerogenes
and envelope toxicity, II: 59
and protein and nucleic acid contents, II: 63
Glycerol 1:2 teichoic acid, II: 44
Glycerol phosphate, II: 42—43
Glycine max, II: 143, 149
Glycogen, I: 116; II: 19, 61
Glycol, I: 100—101
Glycopeptides, II: 50—53
Gonyaulax polyhedra
cell division, II: 178, 179
circadian regulation of, II: 174
Gradient systems, single stream, I: 81—87
Gram-negative bacteria, see also specific organisms
envelopes of, II: 43, 45, 53—55
nutrient limitations and, II: 40, 41
Gram-positive bacteria, see Envelopes, Gram-positive bacteria; specific organisms
Gramicidin-S, I: 104
Grape juice fermentation, I: 82
Growth
balanced, and steady state, I: 144
cellulase-producing system, I: 88
DNA synthesis and, I: 88
in chemostat walls and lines, see also Wall growth, I: 24, 25

in single-stream multi-stage systems, I: 64—75
in tubular fermenters, I: 81—82
phytoplankton
chemostat, II: 166—167
cyclostat, II: 168—169
nutrient concentrations and, II: 162—166
and product formation, in complex systems, I: 49—78
uncoupled, II: 100
with methane/methanol substrates, I: 131—132
Growth conditions
and membrane composition, II: 53
and mutation repair, II: 135, 136
optimizing, I: 107
and selection, II: 137
and yield, on carbon-energy substrates, I: 128
Growth curves, in batch fermentation, I: 98
Growth phases
defined, I: 2—4
and morphology, II: 41—42
Growth kinetics, see Theory
Growth rate
and antibiotic sensitivities, II: 58
and ATP pools, I: 154
and biomass, theory, II: 101
and catabolic enzymes, II: 76
culture conditions and, I: 5—8
dilution rate changes and, I: 14, 30, 145—136
and equations of balance, I: 28—29
and envelope proteins, II: 55
and flow rate, I: 18
gene induction and, I: 159—160
and Gram-positive envelopes, II: 49
and hysteresis effect, II: 127
in methane/methanol based systems, carbon-limitation and, I: 121—122
in single-stream multi-stage systems, I: 66—67, 68
and lipids, II: 55
magnesium requirements, I: 133
and maintenance metabolism, II: 104
maximum, turbidostat culture and, I: 14—15
media and, I: 21
and metabolism, II: 3
minimum, I: 7
and morphology, II: 41—42
and mutation, II: 133
nutrient limitation and, I: 6, 7, 21
and nutrient concentration, I: 124—125
pulse addition of limiting nutrients and, I: 146—149
step changes in limiting nutrient and, I: 149, 151
physiology and, I: 4
phytoplankton, II: 160—161, 162—166, 167
plant cell chemostats, II: 148
and poly-β-hydroxybutyrate, II: 22—24
and protein synthesis, I: 163—164
and respiration, II: 110
and selection, II: 127
and single cell protein production, I: 118—119, 121—123
transients and, I: 159
wall growth and, I: 24
Growth vessel, see also Chemostats; Phytostats; Turbidostats, I: 18—19

H

Haemophilus influenzae, II: 43
Hansenula polymorpha, I: 120
Haplopappus gracilis
growth of, II: 143, 149
metabolite synthesis, II: 143
turbidometry, II: 151
Harrison and Loveless equation, I: 122
Heat
in tower fermenter, I: 86, 87
and yield, in hydrocarbon-utilizing organisms, I: 127, 129—130
with carbon energy substrates, I: 127, 128, 136
Hemagglutinin, envelope and, II: 59
Herbert's theory, I: 88—89
Heterogeneity
plant cell culture, see also Morphology, II: 144—145
Hexose monophosphate pathway, II: 85
High-pressure systems, I: 22
Histidase, I: 160
Histamine sensitizing factor, II: 59
H.R. flow inducer, II: 21
Hydrocarbons, see Carbon-energy limitation; Methane; Methane/methanol
Hydrogen limitation, *alcaligenes eutropus*, II: 19—20
Hydrogen peroxide, and *Lactobacillus plantarum*, I: 153
Hydrogenomonas, polyphosphate of, II: 61
Hydrogenomonas eutropha H16, II: 19—21
Hyperproduction
adaptive significance of, II: 77—78
genetics of, II 129—130, 136
Hyphomicrobium, methanol utilizing, I: 118

I

I.C.I. pressure cycle fermenter, I: 136
Immunological properties, envelope and, II: 59
Induction, see Enzyme induction; Product, induction of
Industrial fermentation, see Fermentation industry
Industrial wastes, see also Waste waters
fixed film fermentation of, I: 102, 109
single cell production, I: 114, 115
Inhibition, see also Catabolite repression
byproduct formation, and equations of balance, I: 29
carbon catabolite repression, I: 153
catabolite in single-stream multi-stage system, I: 73
chemical theory, I: 47—52
gluconate, II: 13

glucose transport, II: 89
methanol, I: 124
of methyl α-glucoside, II: 6
model of, I: 163
phosphate uptake, in phytoplankton, II: 170, 173
poly-β-hydroxybutyrate in *Azotobacter beijerinckii*, II: 31
and selection, II: 133
toxic impurities and, I: 133
and transient responses, I: 152—153, 157
Innoculation
age of culture and, II: 128
contamination during, I: 99
continuous culture, I: 23
and enrichment for hydrocarbon utilizing organisms, I: 116, 117
failure to grow, I: 24
hydrocarbon utilizing organisms, I: 116
plant cell cultures, II: 142
technique of, I: 21
Invertase production, I: 106
Ionic environment, see also Trace elements; specific elements
inorganic ion limitation, I: 132
and gram-positive bacterial envelopes, II: 50
Ipomoea, II: 143
Iron
in fumaric acid production, I: 103
and membrane proteins, II: 55
and pathogenicity, II: 60
Iron limitation, yeast, I: 132
Isochrysis galbana, II: 166—167
Isocitrate dehydrogenase
citrate/glucose ratios and, II: 86
nutrient limitation transitions and, II: 32, 33
Isoenzymes synthesis, II: 89—94
Isopropyl-β-D thiogalactopyranoside, I: 159

K

Karyotypes, plant cell, II: 143
β-Ketothiolase, II: 30, 31, 33, 35
Klebsiella aerogenes, see also *Aerobacter aerogenes*
aerobic to anaerobic transitions, I: 156
2, 3 butylene glycol fermentation of, I: 100, 101
carbon and energy reserves of, II: 17
cofactor pool sizes, I: 155
DNA and protein contents of, II: 62
envelop proteins, II: 55
β-galactosidase, II: 129, 131, 136
induction and catabolite repression in, II: 74—75
low growth rates and, I: 22
low-nutrient environment and, II: 15
nutrient limitation and, I: 6, 7; II: 40—41, 62
parent/variant selection, II: 129
product formation, I: 156
product inhibition, I: 152—153
pullulanase production by, I: 105
slip reactions, II: 109—111
temperature changes and, I: 151
toxicity of, II: 59
xylulose catabolism, II: 78
Krebs cycle enzymes, II: 84—85
Kurz fermenter, II: 144, 145

L

Lac-operon, I: 159; II: 129, 130
B-Lactamase, I: 103
Lactate limitation
and isoenzymes, II: 93, 94
Pseudomonas, and enzyme activity, II: 72—73, 74
Lactic acid
industrial fermentation of, I: 100, 102
limiting nutrient pulses and, I: 147
Lactic dehydrogenase, isoenzymes, II: 93, 94
Lactobacillus arabinosus, II: 42—43, 46, 47
Lactobacillus delbrueckii, I: 102
Lactobacillus plantarum, I: 146, 153
Lactose, and penicillin production, I: 103, 104
Lactose limitation, and β-galactosidase, II: 78, 129, 136
Lag phase, defined, I: 2
Leakage, and maintenance metabolism, II: 119, 121
Liapounov method, I: 35, 36
Light
as mutagen, II: 133, 134, 135
and mutation repair, II: 135, 136, 138
and phytoplankton growth, see Photoperiod
Lipids
and antibiotic resistance, II: 58
Gram-negative envelopes, II: 54—55
nutrient limitation and, II: 53
phytoplankton, circadian rhythms, II: 175
Lipopolysaccharides, phenotype and, II: 54
Lipoproteins, gram-negative bacteria, II: 54
Lipoteichoic acid, in gram-positive bacteria, II: 47—48
Lithodesmium undulatum growth phase, see Exponential growth
Long-term cultivation
and metabolic stability, I: 5—6, 73
and selection, II: 127
Loop fermenters, I: 164
Luciferin/luciferase assay, I: 154
Lymphocytolysis promoting factor, II: 59
Lysozyme sensitivities, II: 50, 51, 59

M

Mackereth oxygen electrode, II: 21, 22
Macrolides, I: 103
Macromolecular composition
energy reserve polymers, II: 60—61
protein and nucleic acid, II: 61—63
Macromolecular synthesis

control of, I: 8—9
in phytoplankton, II: 173
pool sizes and, I: 155
Macromolecules, see also Biopolymers
hysteresis effect, II: 127
intracellular, II: 60—63
nutrient limitations and, II: 40—41
osmotic regulation and, II: 119
potassium and, II: 119
turnover of, II: 118
Magnesium
and ethnolamine phosphate, II: 58
in fumaric acid production, I: 103
and *P. aeruginosa* antibiotic resistance, II: 58
and teichoic acid, II: 45
and teichoic/teichuronic acid content, II: 50
Magnesium limitation
B. subtilis, and phospholipide,II: 63
and cell wall phosphorus, II: 45—46
K. aerogenes, II: 40—41, 59
and lipids, in Gram-negative organisms, II: 54
mixed culture, dilution rate changes and, I: 146
osmotic regulation of, II: 119
P. aeruginosa, and envelope composition, II: 59
and peptidoglycans, II: 50, 51, 52
and plasmid stability, II: 62
and RNA content, II: 62
S. mutans, and morphology, II: 42
and wall composition, II: 48
and wall turnover, II: 45, 46
Magnesium requirements, growth rates and, I: 133
Maintenance energy, defined, see also Bioenergetics, II: 102
Maintenance metabolism, see Metabolism, maintenance
Malt, I: 102
Mandelate limitation, II: 130
Manganese, I: 100
Manganous chloride, II: 134
Mannitol, in mixed culture, I: 146
Mannose, see also Carbon sources
and formate oxidation, II: 102
and maintenance respiration, II: 113—114
and methyl α-glucoside, II: 6
Mannose limitation
formate and, II: 115
and maintenance demands, II: 122
Mashes, alcohol production from, I: 102
Mass transfer, in tower fermentor, I: 86, 97
Mechanical failure, I: 25
Medium , see also Dilution rate
addition of, I: 18
for continuous culture, I: 20—21
for plant cell culture, II: 142
in open systems, I: 1
in tower fermentors, I: 87
methane solubility in, I: 122—123
and product formation, I: 156
spent, collection of, I: 18, 20
and steady state conditions, I: 14
Membrane filtration, I; 21
Membranes, see also Bioenergetics; Envelopes; Transport
and energy spillage, II: 117
osmotic regulation and, II: 119
techniques for separating, II: 53
Metablic activities
antifoam and, I; 24
hysteresis effect, II: 127
in tubular fermentors, I: 81—82
Metabolic coefficient
substrate balance and, I: 72
and substrate consumption, I: 75, 76
Metabolic pathways
location of, in *P. aeruginosa,* II: 11, 12
mutation in, I: 136
pH and, II: 3
and yield, on carbon-energy substrates, I: 128
Metabolic pool sizes, transient phenomena and, I: 154—156
Metabolis quotient
Azotobacter beijerinckii, oxygen and, II: 36
in penicillin production, I: 104
Metabolic rates, plant cells, II: 149
Metabolism, maintenance, II: 117—123
cellular component leakage, II: 119—120
chemical and osmotic regulation, II: 119
growth rate dependent, II: 104
macromolecule turnover and, II: 118
model of, II: 120—123
motility, II: 119
Metabolites, see also Products
and enzyme synthesis, II: 75—77
plant cell, II: 155
toxic, and growth rate retardation, I: 76
Methane
higher alkanes with, I: 136
hydroxylation of, I: 129—130
oxidation of, II: 114
solubility of, I: 131—132
Methane/methanol
affinity of organism for, I: 124—126
biochemistry of utilization, I: 129—131
enrichment for organisms utilizing, I: 116—117
growth physiology, I: 131—132
growth rate with, I: 121—123
mixed vs. pure culture utilizing, I: 118, 120—121
nutrient requirements, 132—133
product production with, I: 115—116
and pure vs. mixed cultures, I: 120
yield coefficients for organisms utilizing, I: 121
Methane mono-oxygenase, I: 129—131, 32
Methanol, in fumaric acid production, I: 103
Methanol dehydrogenase, I: 129, 131, 136
Methanol limitation
and growth rate, I: 118
methanol addition in, I: 132
pulse feeding in, I: 147, 149
Methanol oxidase, of yeasts, I: 115
Methionine limitation
and mutation, II: 133

and RNA synthesis, II: 62
Methylene blue, II: 135
Methyl β-glucoside, and methyl α-glucoside, II: 6
Methylococcus
ammonia transport in, I: 132
carbon excretion by, I: 118, 119
dilution rate in methane-limited culture, I: 125—126
ethane oxidation by, I: 121
growth physiology of, I: 131—132
glycogen-like polymer of, I: 116
substrate affinity of, I: 125
in two-stage system, for SCP production, I: 121
methane oxidation rate, I: 129—132
and nutrient concentration growth rate in, I: 125
pure vs. mixed culture of, I: 118—119
and supernatant carbon of, I: 121
Methylococcus capsulatus, I: 130
Methylomonas methanolica
carbon excretion by, I: 119
productivity of, I: 118
supernatant carbon of, I: 121
yield coefficients, I: 120
Methylosinus trichosporum OB 3b, I: 130
Methylotrophs, nonrespiratory oxygen utilization, II: 114
Methyl viologen, II: 114
Micrococcus glutamicus, I: 103
Micrococcus sodonensis, II: 51
Microcystis
growth rate and subsistance quota, II: 167
nutrient feedback, II: 170
Milk
fermentation of, I: 82
in lactic acid production, I: 102
Mitochondrial respiration, II: 100
Mitotic index, synchronized plant cell cultures, II: 151, 152, 153, 154
Mixed cultures, see also Phytoplankton
algae, II: 175—177
competition in, I: 8
dilution rate changes and, I: 146
for alkane utilization, I: 136
for organic acid production, I: 103
for single cell protein production, I: 118—121
plasmid stability in, II: 62
theory, I: 53—55
transients in, I: 164
Mixed substrate utilization, II: 88—89, 90, 92
Mixing
fermenter design and, I: 133—134
in solid substrate systems, I: 94
in tower fermentors, I: 87
in tubular fermentor, I: 82
methods for, I: 19
plant cell cultures, II: 144, 145
and transient phenomena, I: 164
and yield, with methane/methanol substrate, I: 131—132, 136
Models, enzyme induction, II: 163
Models, growth
Contois, I: 43—44
Monod, see Monod model
structured, I: 44—45
Molasses
for alcohol production, I: 101—102
in glycol formation, I: 100—101
Molecular events, transients and, I: 157
Monochrysis lutherii
chemostat growth rate, II: 166, 167
cyclostat growth phases, II: 164
photoperiodic activities, II: 176
Molybdate, and alginic acid, II: 56
Molybdenum, in butylene glycol fermentation, I: 100
Monascus
enzyme induction in, I: 162—163
β-galactosidase production by, I: 106
Monitoring, culture vessel structure and, I: 19
Monod growth kinetics, I: 30—35; II: 171
extensions of, I: 42—44
growth inhibition kinetics, I: 47—52
inhibitor and inducer effects, I: 163
in Liapounov method, I: 35, 36
methane substrate, I: 124—125
multiple species systems, I: 53—55
multiple substrate kinetics, I: 52—53
phytoplankton growth, II: 162, 163
product formation, I: 80—81
temperature and growth rate, I: 46—47
transient conditions and, I: 145
transients and, I: 39—42
Monomers, leakage of, II: 119—121
MO-104, I: 120
Morinda citrifolia, II: 143
Mucor gerevensis, II: 84
Morphology
bacterial envelope and, II: 41—42
peptidodoglycans and, II: 50
plant cell, II: 144—145
Motility, II: 119
MSI yield coefficients, I: 120
Mucopeptide, II: 50—53
Multi-species systems theory, see also Mixed cultures, I: 53—55
Multi-stage systems
for alcohol, I: 101
for antibiotic production, I: 104
for beer production, I: 108
for bioconversion processes, I: 106—107
for enzyme production, I: 105, 106
for industrial processes, I: 107
for organic acid production, I: 103
holding times in, I: 67, 68, 70, 71
in packed columns, I: 60
position of and gradient systems, I: 82
simulation of plug-flow fermentor, I: 84—85
single stream, I: 64—81
growth characteristics, I: 64—75
product formation, I: 75—81
Multi-stream systems
dialysis feremention, I: 92—93
microorganism inflow, I: 88—92

solid substrate feremementation, I: 93—94
substrate inflow, I: 87—88
Murein, II: 50—53
Mutagens
antimutagen specificity, II: 135
effectiveness of, II: 133, 134—135
Mutants
Bacillus licheniformis, phosphoglucomutase deficient, II: 46
E. coli constitutive for β-galactosidase, II: 77—78
promoter, II: 129
Pseudomonas putida, II: 130
selection for, II: 127—128
wall growth of, II: 132
Mutation
copy error theory of, II: 133
and equations of balance, I: 28, 29
and evolution, II: 136—139
in mixed SCP cultures, I: 119, 120—121
and internal contamination, I: 23
Mutation rate
and evolutionary advantage, II: 138—139
nutrient limitations and, II: 133
Mutator gene, Treffer's, II: 138
Mycelia, in penicillin production, I: 103, 104
Mycobacterium phlei, II: 17

N

NAD cofactors
Azotobacter beijerinckii
carbon metabolism enzymes, II: 28—29
nutrient limitation transitions and, II: 32, 33
oxygen and, II: 28
aerobic/anaerobic transitions and, I: 154
dilution rate and, II: 70, 78—79
in methane mono-oxygenase reactions, I: 129—131
and polyphosphate, II: 60—61
pool sizes, I: 155
1-Napthalenacetic acid (NAA), II: 142
Navicula ostrearia, II: 179
Neomycin resistance *E. coli*, II: 132
Nicotiana tabacum, II: 143
Nicotinic acid, and envelope proteins, II: 59
Nitrates
in phytoplankton, II: 170
waste water, removal of, I: 109
Nitrobacter
growth rate, temperature and, I: 46—47
in packed column system, I: 58—60
Nitrogen, see also Ammonia
dilution rate and, in *P. aeruginosa*, II: 4
energetics of conversion, I: 136—137
for anaerobic cultures, I: 22
for fumaric acid production, I: 103
in alcohol production, I: 101
in high-pressure systems, I: 22
in solid substrate fermentation, I: 93—94
nucleic acid, nutritional value of, I: 122
plant cell cultures synchronizing with, II: 151, 152, 153, 154
Nitrogen fixation by *Azotobacter beijerinckii*, II: 21—22
Nitrogen limitations, see also Nutrient limitation
Azotobacter beijerinckii, I: 22, 23
and oxygen, II: 21, 25, 26
and poly β-hydroxybutyrate regulation, II: 31—32
and supernatant glucose, II: 34
ammonia pulses and, I: 150
Nitrogen limitation
carbohydrate and cryoresistance with, I: 5, 6
and carbon reserve compounds, II: 16
and catabolic enzymes, II: 77
E. coli
dilution rate changes and, I: 145
and glycogen, II: 61
Enterobacter aerogenes and transient phenomena, I: 145
and glucose metabolism, in *P. aeruginosa*, II: 4
K. aerogenes, ethanol addition to, I: 152
pH variations and anionic polymers, II: 50
phytoplankton
and circadian carbon assimilation, II: 174
and RNA and amino acids, II: 173
and polyphosphate, II: 61
and polysaccharide formation, II: 56
and protein synthesis, II: 62
pulse feeding in, I: 147, 149
temperature changes and, I: 151
with phosphate limitation in single-stream multi-stage system, I: 73
Nitrogen/phosphorus ratio and phytoplankton coexistence, II: 175
Nitrogen source
Acer pseudoplatanus, II: 149
in methane/methanol utilizing systems, I: 131
Nitrosomonas
growth rate, temperature and, I: 46—47
in two-stage packed column system I: 60
Nisseria sicca, II: 55
Nitzschia actinasteroides quota, II: 167
Nitzschia turgidula, II: 177, 178
Novobiocin production of, I: 104
Nuclei multinucleated cells, II: 133
Nucleic acids, see also DNA, RNA, II: 61—63
K. aerogenes glycerol-limitation and, II: 63
nutritional value of nitrogen, I: 122
production by *Methylococcus*, I: 118—119
Nucleotides
adenylate dilution rate and, II: 79
pool sizes, I: 155
Nutrient(s), see also Media
addition of limiting, I: 146—149, 151; II: 109
and equations of balance, I: 28, 29
and growth rate, I: 5—8, 99
phytoplankton
and growth rate, II: 162—166
nutrient limited growth, II: 162—169
pools and biosynthesis, II: 173—175
requirements, in single cell protein production,

I: 132—133
shift up of and gene expression, I: 159
supply of in continuous culture, I: 17—18
Nutrient limitation, see also specific nutrients
Azotobacter beijerinckii
and glucose metabolism enzymes, II: 34
and poly-β hydroxybutyrate accumulation, II: 22—24
and poly-β hydroxybutyrate regulation, II: 31—33
transitions of and supernatant glucose, II: 34
Acer pseudoplatanus, II: 149, 150
and anionic polymers, II: 50
and antibiotic sensitivity, II: 58
B. megaterium, and carbon reserve compounds, II: 20
B. subtilis and cell wall, II: 46
and carbon reserve compounds, II: 16
and cell physiology, II: 40—41
criteria for defining, I: 21
E. coli, anaerobic growth of, II: 103
and energy reserve polymers, II: 60—61
and enzyme production, I: 106
growth rate and nutrient concentration in, I: 124—125
and hydrocarbon utilizing organisms, I: 117
limiting nutrients addition, I: 146—149, 151; II: 109
and maintenance metabolism, II: 120
and morphology, II: 41—42
multispecies system, I: 53—54
and mutation selection, II: 138—139
nitrogen and phosphorus, in single-stream multistage system, I: 73
nutrient uptake, feedback, and derivation of Ks, II: 169—173
P. aeruginosa, and glucose metabolism, II: 17, 18, 20
and peptidoglycans, II: 50—53
phytoplankton, II: 160—169
and coexistence, II: 175
and pool sizes, II: 173
plant cell chemostat, II: 148
and reserve biosynthetic capacity, I: 156
and selection, II: 136
and steady state viability, I: 7
and wall growth, I: 24
and wall teichoic/teichuronic acid content, II: 45—47

O

Oligosaccharide-l-phosphate polymers, II: 43
Open systems
classes of, I: 1
plant cell, II: 148—151
Optical density, see Turbidostats
Organic acid, see also Amino acids
and carbohydrate metabolism, in *P. aeruginosa*, II: 3, 4
production of, I: 103
Organic compounds bioconversion of, I: 106—107
Ornithine, and acetylornithinase, II: 131
Osmotic regulation bioenergetics, II: 119
Overflow metabolism, II: 109—111
Oxalacetate, II: 89
Oxalyl CoA reductase, II: 93
Oxidative phosphorylation, see also ATP
and ATP pool size, I: 155
E. coli, II: 103—104, 111—112
efficiency of, II: 107
and respiration, II: 100
and yield shortfall, II: 114—115
Oxidative pathways
in facultative anaerobes, II: 84—85
location of, in *P. aeruginosa*, II: 11
and superoxide dismutase, II: 114—115
2-Oxoglutarate, II: 30
2-Oxoglutarate dehydrogenase, II: 32, 33
2-Oxogluconate
carbon limitation and, II: 14
extracellular pathways of *P. aeruginosa* and, II: 15—16
and glucose transport, II: 10
transport of, II: 11—14
Oxygen, see also Respiration rate
in *Azotobacter beijerinckii*, II: 25—31, 31—33
and enzyme levels, II: 35
and glucose metabolism regulation, II: 33—34
and poly-β-hydroxybutyrate regulation, I: 156; II: 21, 30, 31—32
regulation of enzyme activities, II: 28—29
and tricarboxylic acid cycle, II: 31—33
and ATP pool, I: 154
and cytochrome oxidase, II: 112—113
dissolved, and product formation, I: 156
and *E. coli* enzymes, II: 87
fermentative and oxidative routes in facultative anaerobes, II: 84—85
fermenter design and, I: 133—134
and glucose metabolism, in *E. coli*, II: 87
immobilized cells and, I: 109
in butylene glycol fermentation, I: 101
in high-pressure systems, I: 22
in tower fermentor, I: 87
methane/methanol substrates and, I: 122—123, 130—131, 136
enriched culture, I: 166
heat production, I: 127, 128
nonrespiratory utilization and yield shortfall, II: 111—114
and oxidative phosphorylation, II: 100
ports for measuring, I: 19
and product formation, I: 79
Oxygen limitation
Azotobacter beijerinckii
and product formation, II: 22, 23, 24, 28—29
and supernatant glucose, II: 34
Alcaligenes eutrophus, II: 19, 20
in closed continuous system, II: 148
and maintenance demands, II: 122

and proton extrusion, II: 111
and respiration, II: 113

P

Packed columns, I: 56—60
Paraffins, I: 114, 115
Paraquat, II: 114
Pasteur effect, II: 85—88
Pathogens, culturing, I: 22—23
Pathogenicity, envelopes and, II: 59—60
"Pekilp" process, I: 135
Penicillin production, see also Antibiotics, I: 103—104
Penicillium chrysolgenum, I: 103
Penthachlorophenol, I: 109
Pentose phosphate pathway, II: 149, 150
Peptidoglycans
gram-negative bacteria, II: 54
Gram-positive bacteria, II: 50—53
in envelopes, II: 42
turnover of, II: 118
Peptone medium, and carbohydrate metabolism, II: 3, 4
Perspex sterilization, I: 21
Pesticides bioconversion of, I: 106
Petroselinum hortense, II: 143
pH
fermenter design and, I: 133—134
and Gram-positive bacterial envelopes, II: 50
in open systems, I: 1
in penicillin production, I: 104
and *K. aerogenes* product formation, I: 156
and lactate dehydrogenase species, II: 93, 94
and lipids, II: 53
and metabolic pathways, II: 3
and nutrient precipitation, I: 133
ports for, I: 19
and quality control, I: 99
and xanthan production, II: 57
Phaeodactylum tricornutum, II: 179
Phage, see Bacteriophage
Pharmaceuticals, bioconversion of, I: 106
Phauxostat, II: 113
Phosphate
and alginic acid production, II: 56
in alcohol production, I: 101
esterification of, and oxygen consumption, II: 100
in phytoplankton, II: 170
and lactic acid production, I: 147
and polymyxin sensitivity, II: 58
and respiration in phosphate limited culture, II: 109
and teichoic acid, II: 45
Phosphorus limitation
Acer pseudoplatanus, II: 149—150
Enterobacter aerogenes, and transient phenomena, I: 145
and envelope, II: 45, 59
and lipids, II: 53
and protein, II: 55
and peptidoglycans, II: 50, 51, 52
and phospholipids, II: 54
and polymyxin resistance, II: 58
and wall composition, II: 45, 46, 48, 49
K. aerogenes, and slip reaction, II: 109—110
and maintenance demands, II: 122
and nuclic acid content, II: 62
pH variation and, II: 50
phytoplankton, II: 170—172, 173
and plasmid stability, II: 62
pulse feeding in, I: 147, 149
with nitrogen limitation, in multi-stage system, I: 73
Phosphoenolpyruvate, I: 154
Phosphofructokinase (PFK)
and ATP, I: 155—156; II: 88
Phosphoglucomutase-deficient mutants, II: 46
6-Phosphogluconate dehydrogenase, II: 86
Phosphorus, inorganic
in fumaric acid production, I: 103
in solid substrate ferment ation, I: 93—94
Phosphorus/nitrogen ratio
and phytoplankton coexistence, II: 175
Photosynthesis, circadian rhythms and, II: 174
Photosynthetic purple sulfur bacterias, II: 176
Photoperiod
and coexistence, II: 175—177
and phytoplankton cell division, II: 177—178, 179—181
Physiological cells, I: 4
Physiology
antifooam and, I: 24
environment and, I: 2, 3—4
and single cell production, I: 134—135
with methane/methanol substrates, I: 131—132
Phytoplankton, see also Plant cells
algae, culture principles, II: 160—162
cell division timing, II: 177—178
coexistence and competitive exclusion, II: 175—177
glossary, II: 181—182
nutrient-limited growth, II: 162—169
nutrient pools and biosynthesis, II: 173—175
nutrient uptake, feedback, and derivation of K, II: 169—173
Phytostat, II: 144, 147
Pichia methanotherm, I: 122
Pirt expression, II: 102, 104, 122, 123
Piston flow fermentors, as gradient system, I: 82
Plant cells, see also Phytoplankton
continuous systems, II: 145—153
closed, II: 148
open, II: 148—151
semicontinuous, II: 146—147
synchronous, II: 151—153
culture units, II: 144—145
defined, II: 142
differentiation for product formation, II: 155
genetic stability of, II: 143
suspension cultures, II: 141—143
Plasmids, II: 62

Plug-flow fermentors, I: 81—84
holding times of, I: 69, 70—71
and transient phenomena, I: 164
Polyamines, as antimutagens, II: 135
Polyauxic effect, I: 73
Poly-β-hydroxybutyrate
in *Alcaligines eutrophus*, II: 19—21
in *Azotobacter beijerinckii*, II: 21—34
batch experiments, II: 21
chemostat for, II: 22—24
glucose metabolism enzymes, II: 33—34, 35, 36
limitations affecting, I: 156; II: 22—24, 25—28
oxygen and, II: 25—28, 28—29, 35, 36
regulation of, II: 29—31
tricarboxylic acid cycle regulation, II: 31—33
in *Bacillus megaterium*, II: 19, 20
in single cell protein production, I: 116
Polymers, see also Glycogen; Macromolecules; Poly-β-hydroxybutyrate; Polyphosphate
energy for, II: 105
dialysis culture of, I: 93
and single cell protein production, I: 115—116
Polymyxin resistance, II: 58
Polyphosphate, II: 60—61
Bacillus megaterium and, II: 17
phytoplankton, II: 170, 173, 174
Polyribosomes, see RNA
Polysaccharides
in envelopes, II: 42
K. aerogenes, nutrient limitation and, II: 109
cofactor, I: 157
Pools
and phytoplankton growth rates, II: 166
Population density
determinants of, II: 101—102
and equations of balance, I: 28, 29
Porton Mobile Enclosed Chemostat (POMEC), I: 23
Potassium
energy for transport, II: 108
osmotic regulation and, II: 119
Potassium limitation
Gram-positive vs. Gram-negative organisms, II: 40—41
and peptidoglycans, II: 52
and polyphosphate limitation, II: 61
and wall turnover, II: 48
Predation
in continuous culture, I: 53—55
systems using, I: 91
Pressure hydrostatic, I: 133—134
Product-controlled cultures, turbidostat culture as, I: 16 see also Turbidostats,
Product formation, see also specific products
carbon-energy substrate and, I: 121—123, 128
cellulase, I: 88
closed continuous system and, II: 148
complex systems and, I: 71, 74
dialysis fermentation and, I: 93
and equations of balance, I: 29
fermentation industry
growth associated, I: 100—103
non-growth associated, I: 103—106
transformation, I: 106—107
growth limitation and, in single cell protein production, I: 118—119, 120, 121
induction of, in batch vs. continuous culture, I: 106
inhibitor and, I: 152—153, 157
multiple, I: 106
and packed column dynamics, I: 58, 59—60
plant cells, II: 143, 149, 151, 155
single-stream multi-stage systems and, I: 64—65, 75—81, 84
transients and, I: 156, 164
Product limitation, I: 76, 152—153
Promoter mutation, II: 129
Propane, I: 121, 136
Propionibacterium acidopropionici, I: 103
Propionibacterium shermanii, I: 146
Propionic acid, I: 103, 121
Proteins, see also Enzymes; Single cell proteins
carbon-energy limitation and, I: 124
envelope, II: 42, 54, 55
growth rate and, II: 62
intracellular, II: 61—63
K. aerogenes, glycerol limitation and, II: 63
membrane, immunological properties of, II: 59—60
Methylococcus production of, I: 118—119
turnover of, II: 118, 121
Protein synthesis
Acer pseudoplatanus, II: 149, 150
control of, I: 8—9
deletion mutants, II: 137
and minimum growth rate, I: 7
nutritional shift-ups and, I: 163—164
phytoplankton, circadian rhythms in, II: 175
Q-cells, II: 143
substrate limitation and, I: 73
temperaure changes and, I: 151, 156
transients and, I: 158—159
Protons
extrusion of, II: 111, 114
leakage of, II: 121
osmotic regulation and, II: 119
Pseudomonas
cofactors, II: 78—79
methane utilizing, I: 118
nonrespiratory oxygen utilization, II: 114
RNA of, II: 62
superoptimal temperatures and, I: 10
temperature changes and, I: 151
Pseudomonas C
carbon sources of, I: 136
in mixed vs. pure culture, I: 120—121
Pseudomonas D, yield coefficients, I: 120
Pseudomonas EN, yield, I: 119—120
Pseudomonas aeruginosa
alginic acid in, II: 56
amidase, dilution rate and, II: 76
antibiotic resistance of, II: 58

carbohydrate metabolism in, II: 3—16, 17, 18
carbon limitation and, II: 14—16
citrate and, II: 6
glucose metabolism, II: 6—11, 82—85, 89, 91
nutrient limitation and, II: 18
transport systems, II: 6—11, 11—14, 17
dilution rate and enzyme activity, II: 72—73, 74
envelope composition of, 55, 59
induction and catabolite repression, II: 73—74, 76
phospholipids of, II: 54
Pseudomonas extorquens, I: 124, 134
Pseudomonas fluorescens
lipids of, II: 54, 55
mixed substrate utilization, II: 88, 89
polymyxin resistance of, II: 58
Pseudomonas linderii, I: 154
Pseudomonas methylotropha, I: 147, 149
Pseudomonas NCIB 11264, II: 56
Pseudomonas oxalaticus, II: 93
Pseudomonas putida, II: 130
Pullulanase production, I: 105
Pumps, I: 18, 22
Pyruvate
and methyl α-glucoside transport, II: 6—7
and poly-β-hydroxybutyrate, II: 30
in xanthan, II: 56, 57
Pyruvate dehydrogenase, II: 33

Q

Q-cells, soybean, II: 143

R

Radiation gamma, II: 132
Recycling, see also Biofeedback systems
methanol, I: 124
water, in single cell protein production,I: 136
and yield, I: 132
Reducant limitation, I: 130, 131—132
Reducing agents, in anaerobic cultures, I: 22
Regulation, see Depression; Enzyme induction; Inhibition
chemical and osmotic, II: 119
of product formation, I: 79
repression, II: 3, 129—130
Rel A gene, II: 62
Relaxation time, I: 41, 42
Repression
and hyperproduction, II: 129—130
repressors, II: 3
Reserve biosynthetic capacity, I: 156
Respiration
Azotobacter beijerinckii, oxygen and, II: 36
cAMP and, II: 111
factors affecting, II: 110
growth rate and, I: 7; II: 110
K. aerogenes, slip reactions, II: 109—111
oxidative phosphorylation and, II: 100
oxygen limitation and, II: 113
phytoplankton ,II: 173
with methane/methanol substrates, I: 121—122
Respiratory enzymes
dilution rate and, II: 70
Respiratory enzymes
dilution rate and, II: 70
oxygen and, II: 84
Reversion, *E. coli* trp, II: 131—132
R-factors, II: 62—63
Ribitol phosphate, II: 42—43
Ribitol teichoic acid, II: 43
Ribulose diphosphate (RDP) carboxylase, II: 93
Ribulose monophosphate pathway, in methane/methanol utilization, I: 122, 129—130
Rhizpous, I: 103
Rhizpous nigricans, I: 107
Rhodospirillum rubrum, II: 17
RNA
and amylase production, I: 105
E. coli, nutrient shifts and, I: 149
deletion mutants, I: 37
growth rate and, I: 7
hysteresis effect, II: 127
magnesium limitation and, II: 62
nutrient limitation and, in Gram-positive vs. Gram-negative II: 40, 41
P. aeruginosa, dilution rate and, II: 4
phytoplankton
circadian rhythms, II: 175
nutrient limitation and, II: 173
Q-cells, II: 143
regulation of, II: 62
S. cerevisiae, and growth rate, I: 151
substrate and in multistage systems, I: 88
substrate limitation and, I: 73, II; 62
synthesis of, I: 8—9
temperature and, I: 151, 156
transients and, I: 158—159, 163
turnover of, II: 118, 121
Rosa, I: 143
Routh-Hurwitz criteria, I: 54—55

S

Saccharomyces carlsbergensis, I: 106, 146, 155
Saccharomyces cerevisae
in alcohol production, I: 101
in ATP pool size, I: 154
dilution rate and enzyme synthesis in, II: 81
fermentative enzymes, II: 84
glucose limited, step changes of glucose and, I: 149, 151
pulse addition of limiting nutrients to, I: 146
slip reactions of, II: 111
transient response to dilution rate changes in, I: 145—146
Salmonella typhimurium, II: 118
Sampling, techniques of, I: 19, 21—22

Scendesmus
and *Asterionella formosa*, II: 176
circadium rhythms of, II: 174
cyclostat growth phases, II: 164
growth rate and subsistence quota, II: 167
nitrogen phosphorus ratio for, II: 175
nutrient uptake feedback and derivation of K_s, II: 169
photoperiodic activity of, II: 176
Scendesmus obliquus, II: 179
Scendesmus obtusiusculus, II: 164, 173
Selection, see also Competition; Enrichment; Genetic studies; Mutants; Mutation
factors affecting, II: 127—128
and internal contamination, I: 23
in mixed single cell protein cultures, I: 120—121
parent-variant, II: 128—129, 130
phytoplankton, II: 175—177
plant cell culture, II: 143
in xanthan-producing system, II: 57
wall growth and, I: 24
and yield, on carbon-energy substrates, I: 127—129
Selenastrum, II: 168
Semicontinuous systems, see also Complex systems
algae, II: 163
plant cell, II: 146—147
techniques of, I: 1
Septomyxia affrinis, I: 107
Serine pathway, in methane/methanol based processes, I: 122, 130
Sewage plants
biofeedback system for, I: 91
transient responses in, I: 146—147
Serotyping, techoic acids and, II: 43
Shearing, and viability, I: 7—8; II: 144
Single cell protein
advances in, I: 136—137
continuous culture in, I: 114—115
economics of, I: 122
fermentor design, I: 133—134
future of, I: 134—135
growth rate, I: 121—123
mixed vs. pure cultures for, I: 118—121
nutrient requirements, I: 132—133
organisms for, I: 115—117
substrate affinity, I: 124—126
substrate for, I: 115
Single-stage systems
in enzyme production, I: 105, 106
single stream multi-stage system first stage as, I: 66, 68
substrate utilization, I: 72
with and without biofeedback, I: 89—91
Single-stream systems
gradient, I: 81—87
plug-flow fermentor, simulation of, I: 84—87
tubular fermentor, I: 81—84
multi-stage, 64—81
growth characteristics, I: 64—75
product formation, I: 75—81
Siphon overflow, I: 18
Skeletonema costatum
carbon assimilation, II: 174
cell division, II: 174, 177, 178
cyclostat growth phases, II: 164, 165
growth rate and subsistance quota II: 167
photoperiodic activities, II: 176
slip reactions, II: 100, 109—111
Sodium, osmotic regulation and, II: 119
sodium chloride
and foaming, I: 24
and slip reactions, II: 110—111
Sodium dodecyl sulphate-polyacrylamide gel electrophoresis (SDS-PAGE), II: 54, 55, 59
Solid substrate fermentation, I: 93—94
Soybean, Q-cells of, II: 143
Soybean cells, synchronication, II: 152, 153, 154
Spirillum, II: 93, 94
Sporulating organisms, long-term cultivation of, I: 73
Staphylococcus albus, II: 46
Staphylococcus aureus
cell wall content, II: 46
peptidoglycans, of, II: 51
teichoic acids of, II: 43
Staphylococcus epidermis, II: 46
Staphylococcus lactis, II: 44
Starch, and single cell protein production, I: 114
Starvation, and enzyme induction, II: 130
Stationary phase, defined, I: 4
Steady state
and contamination, I: 99
criteria for, I: 14
defined, I: 4
in biofeedback system, I: 89
in chemostat, plant cells, II: 148
in dialysis fermentation, I: 92—93
in industrial processes, I: 107
in substrate limited growth, I: 30
and long-term cultivation, I: 73
open continuous system and, II: 148
relaxation time and, I: 41
and sampling, I: 21—22
substrate concentration and, 47—52
viability of *Klebsiella aerogenes*, I: 6, 7
vs. balanced growth, I: 144
Sterilization
antifoam, I: 24
techniques for, I: 21
fermentor design and, I: 134
Steroids, bioconversion of, I: 1, 106—107
Stirrers, magnetic, I: 19
Storage
of effluent, I: 22
plant cell lines, II: 143
Streptococcus faecalis
ATP pool size, I: 154
cell wall phosphorus, II: 46
lipoteichoic acids of, II: 48
peptidoglycan turnover, II: 118
Streptococcus mitis, II: 61
Streptococcus mutans

cell wall phosphorus, II: 46
glycogen synthesis, II: 61
limiting nutrient pulses and, I: 147, 148
lipoteichoic acid of, II: 48
membrane composition of, II: 53
morphology, II: 41—42
Streptococcus sanguis, I; 161
Streptomyces erythreus, I: 104—105
Streptomyces aureofaciens, I: 104
Streptomyces niveus, I: 100
Streptomyces 17, I: 160—161
Stress, studies of, I: 9—10
Substrate affinity
in single cell protein production, I: 124—126
and selection, II: 127
Substrate inhibition, models, I: 47—52
Substrate-controlled cultures, I: 16
Substrates, see also Carbon source; Media; specific substrates
alsohol production, I: 101—102
antibiotic production, I: 104—106
for beer production, I: 108
in biofeedback system, I: 90
and biomass theory, II: 101
butylene glycol fermentation, I: 100—101
carbon-energy, see Methane/methanol
dialysis fermentation, I: 92—93
and enzyme synthesis, II: 81—88
in facultative anaerobes, II: 84—85
glucose metabolism, II: 82—83, 84
Pasteur effect, II: 85—88
transients and, I: 156, 157—163
and growth yield, model for, II: 104—109
lactic acid production, I: 102
methanol, toxicity of, I: 124
mixed, and enzyme synthesis, II: 88—89
and glucose metabolism and transport, II: 14—16, 17, 82—83, 84
multiple
in single-stream multiple stage system, I: 73
kinetics, I: 52—53
multi-stage branched system, I: 91
multi-stream systems, I: 87—88
organic acid production, I: 103
penicillin production, I: 103
and product formation in comple x systems, I: 79
single cell protein production, I: 115
single-stream multi-stage systems, I: 64—65, 71—75
solid, fermentation with, I: 93—94
transients in, I: 156, 157—163
Substrate limited growth, see also Nutrient limitation
and product formation, in complex systems, I: 75,'76
substrate affinity and, I: 124—126
theory
chemostat dynamics close to equilibrium, I: 35—39
Monod growth kinetics, I: 30—35
Monod model, extension of, I: 42—44
steady state, I: 30
structured models, I: 44—46
transient behavior, I: 39—42
Succinate
and carbohydrate metabolism, in *P. aeruginosa*, II: 3, 4
and methyl α-glucoside transport, II: 6—7
and nonrespiratory oxygen utilization, II: 115
and oxidative phosphorylation efficiency, II: 107
and respiration, II: 114
and yield shortfall, II: 115
Succinate limitation
and *E. coli* β-galactosidase, II: 75
and maintenance demands, II: 122
P. aeruginosa, induction and catabolite repression, II: 73
vs. carbohydrate limitation, II: 104
Succinic acid production, I: 103
Sucrose limitation, II: 56
Sulfate limitation
E. coli, I: 149
and envelope proteins, II: 55
K. aerogenes, II: 109—111
and maintenance demands, II: 122
and peptidoglycans, II: 51, 52
yeast, and yield, I: 131—132
and yield shortfall, II: 113—114
Sulfate transport, energy for, II: 108
Sulfide limitation, and *Chromatium* competition, II: 176—177
Sulfite
and glucose transport systems, II: 89
waste, and single cell protein production, I: 114, 136
Sulfur, and polyphosphate, II: 60, 61
Superoxide dismutase
dilution rate and, II: 70
E. coli, 114—115
Suspension cultures, plant cell
establishment of, II: 141—143
heterogeneity of, II: 144—145
Synchronication
phytoplankton, II: 160
plant cells, II: 151—153, 154

T

Teichoic acid, II: 42—44, 45—47, 49
Teichuronic acid, II: 44—45, 45—47
Temperature
alcohol production, I: 101
"effective", I; 47
and growth kinetics, I: 46—47, 48
in open systems, I: 1
and lipids, II: 55
monitoring, I: 19
and peptidoglycans, II: 51
and product formation, I: 79
and quality control, I: 99
superoptimal, I: 10

and transients, I: 151—152, 156
Terramycin resistance, II: 132—133
Thalassiosira fluviatilis, II: 166, 167, 179
Thalassiosira pseudonana
cell division of, II: 174, 177, 178
chemostat growth rate equations, II: 166
growth rate and subsistance function, II: 167
nutrient feedback, II: 170
Theory
chemostat
phytoplankton, II: 160—162, 166—169
plant cell, II: 148—151
chemostat, extensions of, 46—60
chemical inhibition and, I: 47—52
heterogeneous conditions, I: 55—60
multiple substrate kinetics, I: 52—53
multi-species systems, I: 53—55
temperature and, I: 46—47
equations of balance, I: 28—30
Herbert's, I: 88—89
maintenance metabolism model, II: 129—130
mutation, II: 133
population density determinants, II: 101—102
substrate affinity, in methane/methanol systems, I: 124—126
substrate dependency, and molar yield, II: 104—109
substrate-limited growth, I: 30—46
chemostat dynamics close to equilibrium, I: 35—39
Monod kinetics, see also Monod growth kinetics, I: 30—35
Monod model, extensions of, I: 42—44
steady state, I: 30
structured models, I: 44—46
temperature, I: 48
transient behavior, I: 39—42
vs. practical values of Y_{ATP}, I: 115—117
Thiobacillus novellus, II: 88, 90, 92
Thiosulfate, II: 89
Time
delays, and Monod model, I: 42—43
holding
in multistage system, I: 67, 68, 70, 71
in plug-flow fermentor, I: 82—93, 84
in simulated plug-flow fermentors, I: 84—86
and product formation, in complex systems, I: 76, 78
relaxation, I: 41, 42
Troula utilis, I: 146
Tower fermentor
design of, I: 86—87
for beer production, I: 108
multicompartment, as gradient system, I: 82
Solid substrate, I: 93—94
Toxins, dialysis culture of, I: 1, 93
Trace substances
and *Cellulomonas* growth, I: 107
in butylene glycol fermentation, I: 100—101
in fumaric acid production, I: 103
and membrane composition, II: 53
toxic, in chemicals, I: 132—133
Transcription, see DNA transcription
Transient phenomena
continuous culture and, I: 5
environmental changes and I: 44—47
aerobic/anaerobic transitions, I: 153—156
carbon catabolite repression, I: 153
inhibitor addition, I: 143
nutritional changes, I: 145—151
temperature changes, I: 48, 151—152
enzyme synthesis during, I: 157—168
glucose concentration, and *P. aeruginosa* carbohydrate metabolizing enzymes, II: 8—9
induction of, I: 144
nitrogen-phosphate limitation and, I: 73
theory, I: 39—42
understanding and applications of, I: 143
Transport systems
anaerobic growth and, in *E. coli*, II: 103
bioenergetics of, see also Bioenergetics
energy for, II: 108
and energy losses, II: 115, 116
and yield, II: 100—101
and maintenance metabolism, II: 120
of growth limiting nutrients, I: 132—133
osmotic regulation and, II: 119
P. aeruginosa carbohydrate metabolism
carbon limitation and, II: 14—16
carbon sources and, II: 9—10
citrate and glucose and, II: 13
and glucose metabolism, II: 3, 4, 6—11, 83, 89, 91
location of, II: 12
methyl α-glucoside, II: 6—7
mixed substrates and, II: 14—16, 17
nutrient limitation and, II: 17
thiosulfate and, II: 89
succinate, energy costs of, II: 115
Treffer's mutator gene, II: 138
Tricarboxylic acid cycle
Azotobacter beijerinckii, II: 31—33
oxygen and, II: 28—29
and poly-β-hydroxybutyrate regulation, II: 30, 31, 32, 33
anaerbiosis and, I: 156
oxygen concentration and, II: 33
P. aeruginosa, II: 3, 4, 6
and polyphosphate, II: 61
Triethylene melamine, II: 134
Triethylene melamine, II: 134
Triticum monococcum, II: 149
Tryptophanase production, I: 106
Tryptophan limitation, II: 133, 134
Tryptophan (tryp) operon, I: 159
Tubular fermentors, I: 81—84
for alcohol production, I: 101
plug-flow, I: 69, 70—71
Turbidostats
absorbance measurements with, I: 19—20
commercially available, I: 25
conversion of chemostat to, I: 20
defined, I: 14—15

K. aerogenes culture in, I: 152—153
nutrient limitation in, I: 21
plant cell, II: 151, 152, 155
selection in, II: 131—132
Two-stage fermentation
antibiotic production, I: 104
enzyme production, I: 105
fumaric acid production, I: 103
for industrial processes, I: 107
plant cell differentiation, II: 155
single cell protein production, I: 121
single-stream multi-stage systems, I: 67, 68, 70
yogurt production, I: 82

U

Uncoupled growth, I: 135; II: 100, 109—111
Ultraviolet light
mutagenicity of, II: 133, 134, 135
and mutation repair, II: 135, 138

V

Vaccines, I: 1, 93
Variants, selection for, II: 127—128
V-fermentor, II: 144, 146
Vicia hajastana, II: 143
Viability
and equations of balance, I: 29
factors affecting, I: 6—8
Viruses, I: 93
Vitamin deficiency, II: 53
Volume
and dilution rate, I: 14
in industrial fermentation, I: 98
and steady state conditions, I: 14
Volutin granules, II: 60

W

Wall growth
and contamination, I: 99
in turbidostat, I: 20
management of, I: 24
mutants and, II: 132, 139
S. mutans, growth rate and, II: 42
and theory, I: 55—56
Walls, cell
and cytoplasmic membrane, II: 42
Gram positive bacteria
teichoic/teichuronic acid content of, II: 45—47
turnover of, II: 48—49
nutrient limitation and, II: 46
Waste waters
biofeedback systems for, I: 91
in solid substrate systems, I: 93
lactic acid production from, I: 102
organic carbon in, I: 120—121, 136
oxygen demand reduction in, I: 109
Weir overflow, I: 18
Whey, I: 102
Williams model, I: 44—46
Wort, I: 108
Wood hydrolysates, see also, Cellulose, I: 100

X

Xanthan production, II: 56—57
Xanthomonas campestris, II: 45, 56
Xanthomonas juglandis, II: 57
"X" mode, II: 59, 60
X-ray photoelectron techniques, II: 45
Xylose, see also Carbon sources, I: 153, II: 6
Xyulose limitation in *K. aerogenes*, see also Carbon limitation, II: 78

Y

Yeast(s), see also Eukaryotes; species
alcohol production, I: 101—102
beer production, I: 108
nutrient limitation, and yield, I: 132
production of, I: 1
single cell protein production, I: 115

Yield, see also Biomass; Cell concentration; Cell mass; Population density
Azotobacter beijerinckii, growth rate and, II: 22, 23, 24
carbon limitation and, see Yield, in single-cell protein production
E. coli, determinants of, II: 103
and economics, I: 107
energy sources and, II: 100
K. aerogenes, growth rate and, I: 7
in single-cell protein production, I: 122, 123, 128—132, 136
carbon limitation vs. carbon excess, I: 147
genetic engineering and, I: 136
heat, oxygen and, I: 127
methane/methanol substrates and, I: 121, 122, 136
mixing and, I: 131—132
product excretion and, I: 118—120
in single-stream multi-stage systems, substrate balance and, I: 71, 72
nutrient limitation and, pulse addition of limiting nutrients and, I: 146—147, 148
nutrient requirements and, I: 132—133
plant cell chemostat, II: 148
and product formation, in complex systems, I: 75, 76
shortfall, II: 109—115
aerobic, nonrespiratory oxygen utilization and, II: 114—115
oxidative phosphorylation and, II: 111—114

slip reactions and, II: 109—111
substrate dependency and, model, II: 104—109
Yogurt production, I: 82

Z

Zinc, I: 100; II: 58